CW00531150

For Tom, Georgie, Gracie and dear Rosie

Getting to grips with end of summer tangles.

Jane Cumberbatch

Pure Style in the Garden

Creating an Outdoor Haven

Pimpernel Press Ltd
www.pimpernelpress.com

Pimpernel Press Limited
www.pimpernelpress.com

Design Vanessa Courtier

A catalogue record for this book is available from the British Library.

Typeset in ITC StoneSerif

ISBN 978-1-910258-06-4

Printed and bound in China by C&C Offset Printing Company Limited

9 8 7 6 5 4 3 2 1

When using the recipes in this book, please be consistent in following either metric or imperial measurements. They are not exact equivalents.

Contents

'There are enough thorns in life, grow a garden.'
Voltaire

As I write, on a dusky autumnal afternoon in south London, my garden continues to nourish as the source of inspiration, beauty, calm, and sanity that it has been through all the lockdowns and restrictions of the Covid pandemic. And with the bulbs bedded down for spring there's a sense of rebirth and possibility, whatever happens next.

This is a book of snapshots, paintings, and notes, from my garden and others which I hope will show you the simple pleasures and constancy in life even when there's so much out of our control. In a world of fake news and virtual communication it shows how the garden can bring back a sense of grounding and reality. It's a book for dipping into again and again or enjoying as one long read (or both).

'The garden is the smallest part of the world and the whole world at the same time.'
Foucault

The garden is a place to be

All gardens hold a universal truth of life and death. And whether it's a single window box with herbs or a bigger space, every garden is grounding. My garden is a retreat as much as it is a place to grow things and interact with nature. The garden speaks to me and is part of my life on many different levels. It is a placing for healing, for staying in touch with the cycle of life; a place to engage with and understand nature and our relationships to it, an outdoor space for living in; and, as an artist, a place to paint.

Carl Jung talks about the collective unconscious, and how the experiences of mankind over millennia have planted the archetype of the garden deep inside us. He believed modern life had separated us from the dark maternal earthy ground of our being. He gardened, and he suggested that we should all have a plot of land so our instincts could come alive again. In tending our plants we're also tending our inner nature.

Nurturing pots of sweet peas mitigated the effects of the severe postnatal depression I suffered. And some years later, when my marriage ended, the garden again proved itself as a safe place, where I felt a sense of purpose and positivity. Gardening is a shock absorber for tough times: when you work with your hands, weeding or snipping, you free your mind to work through feelings and problems. Often if I'm feeling upset I go and have a good hack at a bramble, or tidy a rose. Seeking for a gentler and less confrontational place to break the news that my mother was dying, my father decided to tell me as we were digging up potatoes together in the vegetable patch. I recently caught an uplifting documentary series, *The Gardeners of Kabul*, about Afghans who find solace in the private gardens of Kabul as they lovingly tend flowers and vegetables, using the process as a form of therapy and escapism. While moving a potted plant in his floral patch, eighteen-year-old Hamidullah says, 'Gardening is a kind of temporary peace for the people.'

Being in the garden makes you look more closely at the world, the way the petals are formed on a flower or how a spider spins its web. I can't help but anthropomorphize my garden. I even feel that I owe it to the garden to look after and care for it and all the things that flower in it. It gives me so much pleasure and my care is reciprocated in the beautiful scents and colours and textures. I have become so attached to the dear roses, for example, that they're like old friends and in turn it's a joy to share them in cut bunches for my other friends. As for the apple tree, when part of it collapsed a few years ago in a wind I was in floods, as if someone had died. (PS It has carried on growing and flowering.)

I come from a line of everyday gardeners. My maternal grandfather, an unreadable man and owner of a small hardware shop in Clapham, south London, which was bombed out during the Second World War, poured his fury

into a pocket-handkerchief garden where he had a way with raspberries, dahlias and a small rock pool decorated with water plants. His skill in taking cuttings of scrappy twig and magicking new shoots was inherited by my mother. She potted up nascent lavender, geraniums, or whatever, on the sunny kitchen windowsill, with a nurturing eye on their progress over the washing-up. My paternal grandmother and her sisters kept the large shaggy front lawn vaguely in check at their Victorian seaside villa home in Torquay, Devon. It was as much as they could do. During the war, the small backyard became a Dig For Victory patch with potatoes and runner beans made fat by trenches lined with seaweed that my teenage father carried in sopping buckets from the beach.

My garden passions were rooted at home, a 1930s semi in the south London suburbs of the 1960s, where my mother created a haven of lawn, climbing roses, and herbaceous borders with stocks, delphiniums, pepper- and clove-scented pinks. One summer she gave me a small patch in which to plant annuals – candytuft and marigolds – and kick-started my passion for gardening.

Despite its limited size our garden had a dimension of freedom and exploration that playing indoors didn't. In the long summer holidays my sister and I ran barefoot and built tented retreats out of deckchairs and blankets in which to read our library books. Or we squeezed into the leafy cave between the lilac and the wooden fence and held dolls' tea parties with daisy cakes and grass tea. A swing was simple transport that took us to a bigger space, back and forth, higher and higher into reaching blue suburban sky.

We buried guinea pigs Miss Muffet and Little Black Sambo (the much-loved central character from our book of the same name – in those days racial insensitivities were yet to be acknowledged) in a rose petal grave under the apple tree. They joined the matchbox tombs of a queen bee killed by the cold and a fledgling thrush tipped out of a nest in the tree. Sometimes we would dig them up, curious to see the process of desiccation: part of the cycle of life and death within the dying back and renewal of the garden itself.

Summer days in the garden with my sister, Sarah.

When my GP father was promoted to senior partner in his medical practice, we upgraded to a larger detached 1930s house, with a more expansive garden and the added excitements of a small summer house and two large Victoria plum trees. My mother grew swathes of sweet yellow 'Peace' roses, and planted tomatoes, courgettes and raspberries between the flowers. I sunbathed on the lawn, a calming retreat from boyfriend problems and the mysteries of O-Level maths (three attempts to get it). If not particularly willing (I did it only because I was asked to), I secretly came to enjoy watering the borders, deadheading roses, and mowing the grass – repetitive gardening tasks that created headspace during the seemingly never-ending summer seasons of exams.

Forming ideas

If my childhood gardens are blueprints, then the osmosis of ideas through my work as a stylist and designer, and the books and columns I've read on gardens in history and literature, have all helped in forming the garden between these pages as a calm retreat and place of beauty at every time of year. I don't dream of some unattainable outdoor utopia, ever-flowering, ever slug-free. I have crafted my garden to fit my personal notions of what is sensuous, pleasurable and simple, without crazy expectations and limitations.

I am not interested primarily in the garden value of a plant, or if it is in or out of fashion, but rather in the intrinsic beauty of its flower, its proportions and shape, and in contrasts, of the relation of the leaf to the bloom or the plant to its neighbour. I'm beguiled by the changing seasons. Spring doesn't mark the beginning or autumn the decline of every herb and flower that grows, for plants are springing and fading continuously throughout the year. I can walk through the garden in bleak midwinter and see buds on bare branches, themselves sappy and green inside; dangling seed pods on desiccated stems from the autumn and leaves both evergreen and not. There are flowers too, tiny cobalt-blue petals on the rosemary, a lingering rose bloom, and the new yellow buds of narcissi pushing up. Then in late spring, when nearly everything is bursting with freshness, youth and newness, the mature apple blossoms are falling and potential fruits swelling in their place, the last tulips are elegantly dropping, while the dahlias are only just sending out sturdier leaves and shoots.

What makes a beautiful garden? A neglected wilderness has its own beauty – knee-high grass, a few twisted fruit trees and a fence overrun with tumbling nasturtiums – until it is ordered and then the loveliness is gone. On the other hand, an exquisite zen-like space with a perfectly formed tree and a few perfectly formed plants is limiting if one wants to add more informality. I am saying that my way of gardening is a kind of middle route (although not middle of the road in the least, I would hope), where irregularity works against and as a foil for regularity – balance and harmony, perhaps.

Having a go at gardening

If I count the starter flower patch of childhood, this garden is my sixth. The second, when I was in my twenties and just married, was a small square affair behind a Victorian terraced house off Columbia Road, home to the Sunday Flower Market in Bethnal Green, east London. Knockdown-priced plants as the market closed provided a cheap way to experiment, and soon climbing pink and white geraniums and other delights spilled over pots and troughs. I also managed to revive an ancient rose against the back garden wall: it grew out of a earthy rubbish mound filled with the remains of hundreds of oyster shells, apparently as everyday as fried eggs to the Victorian Londoner.

We were suckers for building challenges. Our next move was to restore a derelict early eighteenth -century house not far away in quietly decaying and neglected Spitalfields (and next to the eponymous flower and vegetable market) with a damp, dark high-walled yard which looked on to the neighbour's outdoor lavatory arrangements. 'Madame Alfred Carrière', a white scented rose which doesn't mind north-facing aspects, together with white *Clematis montana* and

C. 'Henryi', which also flowered in the challenging conditions, helped make the most of the dingy proportions. In addition, we could also escape to a small sunny roof terrace and although this meant copious watering on hot days, I crammed it with herbs, sweet peas and more clematis.

Garden number four continued the courtyard theme, but in a more extreme climate at our home high up in the Andalusian sierra where David Austin white roses (brought out in a suitcase) and white jasmine flowered profusely around whitewashed walls. I also discovered that the robust pink geranium that everyone grew in the village provided perfect leafy texture, colour and lemony scent: a case of mixing the imported and the local to make the space feel more timeless.

After the Spanish adventure, we arrived at garden five, the minimal whitewashed courtyard at the house in Olhão, Portugal (we still have it, and love it), with a simple arrangement of three large terracotta pots planted with purple agapanthus – beautiful when the purple spikes on long elegant stems are in flower in high summer, and also afterwards when the dried stems make architectural plant detail in winter. These are very special plants – given to my daughter on her eighth birthday, surviving various dividing-ups and a move from southern Spain. I water when I visit, and friends come in at other times.

My first encounter with the garden which takes up most of the following pages was seventeen years ago on a late summer morning of heat haze thick with clouds of flying ants. It was forlorn and overgrown. Dense banks of brambles choked the shrivelled remnants of once-cared-for shrubs. The lawn was rampant and high, wild and gone to seed. The tattered branches of an apple tree trailed through the grass thickets. The flower beds were furred over with grass and more brambles. I traced the ghost of a garden path running past skeletal rose bushes starved of food and attention and through smeary shed windows saw a hunched group of old hand lawn mowers with rusted blades.

Elderly Mr Campbell, who had lived here since birth, had recently died and cousins were selling the Victorian house his parents had bought in 1896, when it was newly built on the remnants of old orchards in Tulse Hill, south London. Eight months later, after the ups and downs of buying and selling, we picnicked and planned among the brambles and spring bluebells excited to be the new custodians of the house and garden.

Laying it out

Summing up my various garden encounters over the years, I guess the running theme is that I try to make the best of and work with what I have. Perhaps it sounds pretentious to say that I hope I am in the tradition of Alexander Pope, who in the eighteenth century advised his client the Earl of Burlington that the gardener should follow nature and 'consult the genius of the place in all'. But that's how I feel.

The garden is 22 metres/70 feet long by 11 metres/36 feet wide and north-facing. There are three sections: The kitchen door leads down steps to a patio laid with worn and weathered reclaimed Welsh bricks – a space for eating from which you can view the whole garden. The middle section is a flower and herb plot bordered by two central metal 'gothick' arches. I have always been charmed by old-fashioned potagers and parterres and I dug out sixteen rectangular beds lined with cobbled edges saved from the original garden and divided by gravel paths.

This is a little world within a world, a place to wander among the tulips in spring and in summer to brush against scented lavender spilling over the paths, with the hum of bees all around. Then there is the grass area (not a lawn I should add – I'm not really a lawn person and am quite content to share my grass with clover and dandelions) beneath the spreading branches of the apple tree. I let the grass grow high for the resonance it has of a country orchard. My parents' wooden bench seat by the shed in the bottom corner is the perfect place to soak up the last of the sun's rays before it sets below the rooflines.

The garden view in spring.

Into the Cottage Garden

I am a city girl by birth but my heart is in the country – or at least in the idea of the country. I love and need nature but find it hard to live without the urban buzz. It's a happy compromise therefore for me to live in the busy vibe of south London while realizing my vision of a garden that contains the elements, spirit and feeling of a cottage garden in the countryside. That goes, too, for the way I decorate the house, with natural textures, and flowers and herbs gathered from the garden.

Cottage gardeners

When I first saw the White Garden that novelist, poet and gardening writer Vita Sackville-West made at Sissinghurst (not just white, by the way, but layers of whites, silvery greys and greens), I was beguiled – enchanted by its controlled chaos and abundance of flowers and herbs in a series of garden rooms bordered by topiary box hedges and weathered brick paths. I later read that Vita was inspired by Gertrude Jekyll, revitalizer of the traditional cottage garden which had fallen out of fashion in the nineteenth century. 'I have learnt from the little cottage Gardens that help to make our English wayside the prettiest in the temperate world. One can hardly go into the smallest cottage garden without learning or observing something new,' wrote Jekyll. Planted with an artist's eye for colour and texture, her naturalistic creations were sophisticated but had the spirit of their more modest inspiration. 'The size of a garden has very little to do with its merit. It is merely an accident relating to its owner. It is the size of his heart and brain and good will that will make his garden either delightful or dull.' I'm with her there.

The first cottage gardens had little room for decorative plants, they had to provide for the family: pigs and chickens would have shared the space with the plants grown to feed them. Staple vegetables were cabbages, broad beans, carrrots, kale, leeks, onions, parsnips, and – from the eighteenth century on – potatoes.

Fruits were apples, damsons, cherries, gooseberries. And then there were herbs for flavouring and medicinal use: borage, chives, dill, garlic, horseradish, fennel, hyssop, mint, rosemary, sage, sorrel and thyme. Flowers were often interlopers from the gardens of monasteries and great houses.

I am inspired by the simple natural gardens I come across on my wanderings (*see opposite*). I also hold a special place in my gardening heart for Manor Garden allotments, tragically bulldozed to make way for the 2012 Olympic site, where an old friend, John Matheson, worked a magical vegetable and flower patch among the communal gardens on a breezy hill. An urban oasis wedged between pylons, scrapyards and ruined factories, with a view of a sliver of a tributary of the Thames, where herons dived and frilly cow parsley grew head-high in summer. Only those with highly detailed instructions and a key were able to access this little inner city haven where retired publicans and railworkers tended their dahlias, fed the foxes and offered gardening advice alongside families with young children who spent weekends weeding, planting and harvesting home-grown produce. On one side of a thick rosemary hedge the flower garden was planted with alliums, euphorbias and lavender. Six types of potatoes, pumpkins, French beans, rocket, courgettes and other delicious crops appeared through the growing season. There was a toolshed for stowing garden kit and deckchairs.

Even if we have lost that particular haven, it is good to see now that there are many new initiatives to bring greening to the urban landscape: guerilla-gardened flower beds in high streets and urban orchards are on the rise. And in the streets around me I notice the proliferation of endearing front gardens and window boxes planted up with alliums and all manner of wild flowers, by a new generation of young and passionate gardeners working from home but needing connection with nature and respite from online work.

TOP TO BOTTOM, LEFT TO RIGHT: A cottage garden on Seil island off the west coast of Scotland; garden at Charleston, East Sussex; Gertrude Jekyll in her gardening hat; a wigwam of plant supports, Charleston; *English Cottage Gardens* a favourite book by Edward Hyams; Vanessa de Lisle's pretty London garden shed; a flint wall in Southwold, Suffolk; a postcard of *Le Jardin* by Pierre Bonnard; front garden of a period cottage, Melbourne; Kim Cargill's lavender garden in south London; the garden at Bloom (decorative antiques), Sag Harbour, New York; hollyhocks in a Southwold garden; elegant irises in an east London urban front garden; greenhouse at Charleston; traditional red brick path detail, east London.

English
Cottage Gardens

Garden Ingredients

This isn't a definitive list of plants and flowers but those below are the ones that I know and love, and grow for their particular features – colour, scent, texture, length of flowering, even all these elements combined – which contribute to the overall look and feel of my garden:

HERBS

Allium schoenoprasum	Chives
Anethum graveolens	Dill
Eruca sativa	Rocket
Laurus nobilis	Bay
Lavandula angustifolia	Lavender
Melissa officinalis	Lemon balm
Mentha spicata	Spearmint
Mentha × piperita	Peppermint
Origanum majorana	Marjoram
Origanum vulgare	Oregano
Salvia officinalis	Sage
Salvia rosmarinus	Rosemary
Thymus	Thyme
Tropaeolum	Nasturtium

Rosa **ROSE**

'Constance Spry' · 'Crocus Rose' · 'Eglantine' ·
'Gertrude Jekyll' · 'Grace' · 'Iceberg' · 'John Clare' ·
'Kiftsgate' · 'Madame Alfred Carrière' · 'Mary Rose' ·
'Saint Swithun' · 'Teasing Georgia' · 'Winchester Cathedral' ·

Allium **ALLIUM**

'Ambassador' · *angulosum* · *cristophii* · 'Firmament' ·
'Gladiator' · 'Globemaster' · 'Mont Blanc' ·

Dahlia **DAHLIA**

'Franz Kafka' · 'Leila Savanna Rose' · 'Onesta'

Thistle, roses and verbena from the summer garden, a watercolour painting.

Tulipa TULIP

'Bleu Aimable' late season · 'Blue Heron' mid-season · 'Blue Parrot' mid-season · 'Purple Dream' mid-season · 'Recreado' late season · 'Rems Favourite' mid-season ·

AS WELL AS:

Buxus	Box
Centaurea cyanus	Cornflower
Cynara cardunculus	Cardoon
Echinops ritro	Globe thistle
Fragaria vesca	Wild strawberry
Hyacinthoides non-scripta	Bluebell
Malus domestica	Apple
Myosotis	Forget-me-not
Narcissus	Daffodil
Nigella	Love-in-a-mist
Paeonia	Peony
Papaver	Poppy
Polygonaceae	Rhubarb
Verbena bonariensis	Verbena

Winter

Things never quite shut down in the winter garden. Fresh beginnings are quietly stirring beneath the dark earth and in the xylem and phloem of roots and stems.

Winter garden surprise: ice crystals decorate a rose hip like a layer of sugar frosting.

It sifts from leaden Sieves
It powders all the wood
It fills with alabaster wool
The wrinkles of the road . . .

Emily Dickinson, 'The Snow'

With the last dregs of leaves fallen, garden colour switches to low volume (save for rich red 'Kiftsgate' hips and some rose bloom bursts); and then I feel it's winter, or at least more winter than any lingering attempt at being autumn.

Like the furring of borders between countries where language and culture cross over, winter and autumn – and, indeed, summer and spring – are marked by some influence of their adjoining seasons. More so in temperate climates, where sometimes our mellow winters feel more like autumn or spring. I continue to be seduced by the prospect of a winter wonderland – that is, until the heating seizes up and the gas supply crashes. (It's the reverse in a heatwave summer, when we fling our cares away in sun-baked parks, but need copious water supplies in order not to die on buses or tubes.)

From the train on the way to see my sister in Somerset, the stark silhouettes of trees and chalky fields with scatterings of birds over Salisbury Plain fuel my sense of winter. The shortest day, on 21 December, is something of a hurdle that once reached fills me with optimism that the grim dark will be lifting in small daily increments. And then there's Christmas, and all the jollity of being inside with sweet orange and pine-needle scents.

As the garden slumbers on without the frenzy of spring and summer one can linger over tasks like washing down outdoor furniture. From midwinter, rose pruning can be attended to – in my case intermittently, between indoor duties, and whenever I need some outdoor inspiration and exercise.

The basics of rose pruning are to cut out dead and diseased stems, spindly and crossing stems. Aim for well-spaced stems that allow free air flow. On established roses cut out poorly flowering old wood and saw away old stubs that have failed to produce new shoots.

To keep my flowering shrub roses, like 'Gertrude Jekyll' and 'John Clare', healthy and in check, I cut back by up to a third in late winter, and take some of the older main stems back to the base.

I grow dear 'Constance Spry' as a climber and in late winter give her the rose version of a short back and sides, removing all dead and straggly stems and cutting back side-shoots to two or three buds. I think I might have gone too far when I see her shorn appearance, but the mass of blowsy pink blooms each summer happily proves otherwise. Similarly, the spreading branches of Rambler rose 'Kiftsgate' (like 'Constance' flowering only once) have a rigorous cut-back. I also remove several of the oldest stems back to their base, to keep this wild child in check.

If you're beginning a rose adventure in the garden, and have ordered a bare - root rose (a dormant plant without leaves) soak the roots in a bucket of water for at least two hours. Dig a hole 30–45 cm/12–18 inches deep and 60 cm/24 inches wide. Back -fill the planting hole with water and allow to drain. Layer with compost or well-rotted manure.

Winter pruning for the apple tree is aimed first at cutting out diseased, crossing and upright branches, then at thinning the canopy and removing any new growth which does not spread away from the centre of the tree. Always cut above a bud that's aimed outwards.

Winter pickings: cardoon leaf, verbena,
rose with hips.

11 December
It's getting seasonal. Last night, before leaving Olhão for London, I grabbed a steaming brown paper bag of hot chestnuts (castanhas) laced with a pinch of rough sea salt, from the seller who roasts them in glowing coals from a rickety two-wheeled wagon under one of the jacarandas in the Avenida. I also packed Christmas treats of fig cakes decorated with almonds, and wedges of dried tuna (mojama).

The garden is going into dormancy, all is quiet and dying back; I can see right through the Norway maple now and to the detail in the gardens backing on to mine. It's so uplifting that the roses continue with almost defiant punches of colour and scent.

So enjoying Sue Stuart-Smith's, *The Well Gardened Mind*, in which she investigates the positive effects of the garden and nature on our health and wellbeing.

'John Clare' roses are wonderfully defiant in the way that they bloom even through the frosts.

18 December

I like to get warm and close up with a wood-burning stove, a back-to-the-fire -in -the-cave thing. I admit I've associated wood-burners with being rather benign . . . in that kind of magic-thinking way that leads us to believe that the things we want for ourselves are harmless, even moral and essential. But some research has shown that between a quarter and a third of all London's fine-particle pollution comes from domestic wood-burning. I have stuck to the regulations by installing a modern stove with low smoke emission when used with wood that's dry and ready to burn, but somehow I feel that my wood-burning days in the city should be numbered.

Our home-grown wreath of yew, bay, ivy and rose hips.

20 December

Remember the cultural moment that Danish *hygge* had a few years ago, when winter was all about cosy candles, fireplaces and blankets? And then there is the Norwegian mindset of *koselig,* similar to *hygge* but with special emphasis on a kind of shared, safe togetherness including connecting with others and spending time out in nature. A recent study on winter mindset by psychologist Kari Leibowitz found that the further north people live in Norway the more positive they tend to feel about winter (making the best of a bad deal?). I embrace the relatively moderate English winter, with moody flickers of candlelight, and cuttings of bay and ivy strewn across the mantelpiece and tucked above the mirror. I cut bay, rosemary and yew to make simple seasonal wreaths, symbols of growth and the cycle of life that involve little more than wire bent into a circle (a coathanger will do), around which you weave the twiggy greenery, tying it in at the edges with matching green string.

I cut stems and sprigs from the garden for natural seasonal decorations.

27 December

Long rays of afternoon winter sun flicker across the walls, bathing them in warm light. There's a lemon-sherbety whiff of summer from the garland I've strung up for Christmas, a winter version of the autumn leafy one, using trimmings from the garden – agapanthus heads, lemon balm and bay sprigs – and a spray of seasonal mistletoe from the market.

Some people really don't like all-white schemes, and indeed they can be particularly relentless on the eye, especially the minimalist type where the stuff of everyday life is hidden away behind cleverly designed white fixtures and fittings. White works for me when it's combined with natural textures and colours that play to its whiteness, just as the blues of sea and sky, and the greens and earth colours of landscape do to white puffy clouds. Covering the big sofa in blue-and-white-striped cotton and painting the floors and walls white is my way of bringing inside the feeling of summer coastlines, light and freedom.

Natural textures: a hyacinth in a terracotta pot, with beach pebbles at the side.

30 December

I empty an almost-full wheelbarrow of rainwater. Then I park the tin bath planted with white 'Thalia' narcissus under a garden table: the rains have been so torrential that I fear bulb rot.

Natural details: a winter garland of garden ingredients above a sofa covered with blue-and-white-striped cotton.

In Touch with Nature

I'm intuitively drawn to furnishing my life with natural materials, colours, textures, sensations, and relics of nature itself. I'd rather sleep in a field than between nylon sheets. This might sound overly fussy, but, far from being precious, it seems I am in tune with my hunter-gatherer DNA . . . and 'biophilia', a term first proposed by German psychoanalyst Erich Fromm in the 1960s to describe 'the passionate love of life and all that is alive . . . the wish to further growth, whether in a person, a plant, an idea or social group'. In the 1990s biologist Edward O. Wilson goes further and describes 'biophilia' in his eponymous book as humanity's innate affinity for the natural world – which was, of course, the main influence on our evolution, on our cognitive and emotional learning. It helps explain why we are captivated by blazing fires; why a garden view can enhance our creativity; why shadows instil fascination and fear, and why the presence of the family dog, cat or hamster promotes healing and restoration.

As modern stresses burden the human psyche the concept of 'biophilic design' is a growing theme: the point being to increase well-being through both direct and indirect contact with nature. Look at Face Book's pre-pandemic rooftop garden at its Silicon Valley hub, and spherical conservatories at its Seattle headquarters. And look, too, at the way the new default of 'working from home' has encouraged us all to imbue our work space with plants. Jana Soderland, author of *The Emergence of Biophilic Design,* explains that biophilia is about bringing into our immediate surroundings nature in all of its forms 'including patterns, shapes,spaces,smells,sights and sounds'. Natural light supports the circadian rhythms of the body which regulate our sleep/wake cycles, as well as our hormones. Natural objects such as seashells or stones reinforce our senses.

There's nothing new about 'biophilic design'. Nature themes are found throughout the history of architecture, from Stone Age huts to the courtyard gardens of the Alhambra. Connections with nature are vital for living. Since the mid-nineteenth century we have all been influenced by the 'biophilic design' of the Arts and Crafts Movement. 'Have nothing in your home that you do not know to be useful or believe to be beautiful,' said William Morris who followed John Ruskin in believing that society should work towards promoting the well-being of its members by creating a union of art and labour.

'I do not mean to assert that every happy arrangement of line is directly suggested by a natural object; but that all beautiful lines are adaptations of those which are commonest in the external creation . . . The pointed arch is beautiful; it is the termination of every leaf that shakes in summer wind and its most fortunate associations are directly borrowed from the trefoil grass of the field or from the stars.'

John Ruskin, *Seven Lamps of Architecture*

William Robinson, gardener and journalist, collaborated on many of Gertrude Jekyll's cottage-garden-inspired projects, using natural materials, waving paths, wild flowers and herbs. Similarly, American architect Frank Lloyd Wright abstracted prairie flowers and plants for his windows and ornamentation, and opened up interior spaces that also revealed views with intimate refuges. Later, Le Corbusier's unrealized Cité Radiant urban masterplan in 1924 was designed to contain high-density skyscrapers spread across a vast green area, allowing the city to function as a 'living thing'. (We are still unpicking the results of postwar high-rise estates ostensibly inspired by Le Corbusier but which failed to recognize the 'living thing' aspect.)

Winter whites: simple pull-on cotton cover for a junk chair and dried cardoons with textural detail.

1 January
The inaction of the New Year garden seems to reflect its indifference and mine to the hype of January fitness regimes and exclusion diets. More interestingly, I am thrilled to see the apple tree bud with furry nodes at the tips of stark branches: harbingers of spring, positivity and renewal.

It's convenient to ignore the garden's winter déshabillé when the daylight hours are short. 'Good, let it stay like that until spring,' is cheering advice from Alys Fowler in *The Guardian*, urging us not to touch our gardens over winter, 'so that's there a place to hibernate for butterflies and other insects'.

4 January
Wrapped in its winter blanket, the garden palette still has a subtle dynamism changing from day to day – even hour to hour. We wake to a gothic winter fairytale garden wreathed in ghostly white veils of freezing fog and a white sprinkling of ice clinging to the cardoon seed heads.

By lunchtime the fog has dissipated, flint-grey cloud has moved in and the scene has turned to a two-dimensional canvas of matt browns and greens.

The garden veiled in ghostly freezing fog.

7 January
A damp and humid thaw moved in overnight. The trees drip, drip, and beads of water soak my trainers en route to the compost bin behind the hazel fence. Dusk is falling even earlier today because of the cloud thickness.

The patch of rhubarb by the kitchen door is shut down for winter, no sign yet of the shiny pink crown buds that will unfurl into a lush spreading plant with leaves the size of tennis racquet heads.

For now, there's the mood-lifting visual and taste pick-me -up of vivid fuchsia-pink forced rhubarb to be had in the shops, sourced from long, dark barns across a small patch of West Yorkshire where a long agricultural tradition continues.

This is the rhubarb triangle, between Wakefield, Morley and Rothwell. Here, from January to March, forced rhubarb, prized for its subtle flavour, is picked by hand, by candlelight so that the delicate stems are not turned green and hard by photosynthesis. I have to recommend a salad of thin slices of raw forced rhubarb together with fennel and carrot and a lemony dressing – all pink, white and green and very fresh-looking.

Vivid pink hues of forced rhubarb make it perfect watercolour material.

10 January
Low afternoon sunlight floods the brick path, outlining dear Rosie's black furriness in liquid silver. It will be freezing up again and so we have to go out for some bramble and weed extraction before the wind blows in from the Arctic.

LEFT: Rosie sniffing the breeze on a sunny day in the winter garden.
OVERLEAF: Scribblings and garden nature details from my January notebook.

↑
Cuttings from
the January
garden: rosehips,
last few Winc-
hester Cathedral
roses, verbena,
spiky green
cardoon leaves

Some things to do
in January
Order Compost, seeds, rose-fed
Mend broken tools, eg. rake;
Sharpen Shears
Deadhead tops of non-shrubby
plants, unless for decoration
during these flower ad
colours less months. e.g.
Cardoons ad Verbena.
Can sow broad beans for
an early crop - sow in 2 rows
7 inches apart. and allow
6 inches between the seeds
in each row.
ing for compost

More Winter Pleasures with Old Friends

Aside from planning and dreaming vicariously through vibrantly illustrated plant and seed catalogues, winter is a time for retreating by the fire or to bed with old friends from my bookshelves.

Classic food writer Elizabeth David's simplicity of expression 'the ever-recurring elements in the food through these countries are the oil, the saffron, the garlic, the pungent local wines; the aromatic perfume of rosemary, wild marjoram, and basil drying in the kitchens' (*Mediterranean Food*, 1950) is echoed in my also very well thumbed *Garden Design* (1982) by the late and great interior decorator David Hicks. It is wonderfully practical yet visually inspiring, even in black and white, with photography by the author. To be sure, it features gardens with ha-has and grand sweeps, but Hicks explains that his ideas for landscaping and good-quality material – natural wood, flagstones, yew hedges, simple garden furniture – are universal and can be adapted to any garden,large or small.

Another of my timeless go-to sources of wisdom is the green-linen-bound *Outlines of a Small Garden* (C. H. Middleton, 1945), picked up in my local charity shop. Cecil Henry Middleton or 'Mr Middleton', as he was known by his listeners was a kind of 'Gardening Treasure' of the 1930s and 1940s, when he broadcast for the wartime 'Dig for Victory' campaign. His approach was direct and reassuring, no doubt a comfort in the uncertain times. 'I'm afraid I am rather unorthodox. So long as a plant looks nice and grows well, I never care a rap whether it is in its so-called correct group or not. The result is that my borders include herbaceous and other perennials, biennials, annuals and even shrubs, all growing happily together, and there is usually a good show of flowers of one kind or another.' Or, on making a potato 'pie' or 'clamp', 'an easy, economical and quite satisfactory method of keeping potatoes'.

I'm so glad I didn't throw out my early copies of *Gardens Illustrated*. It is so beautifully spare in design, with great photography and good paper, and it covers a wonderful diversity of gardening themes, which continue to give pleasure and to inform: May 1999 features umbellifers (the plants with upside-down umbrella shapes) with hero portraits of angelica and a painterly soft -focus close-up of frothy fennel. In the same issue seventy-six-year-old twins Alf and Jim Howe's allotment is a hotbed of activity: 'Our father used to say that when you were old enough to sit on the ground you could plant in it.' In March 2000 it's fascinating to see Dream Park, a public park in Sweden and an early Piet Oudolf collaboration, with his trademark grasses and perennials.

It is good to be immersed in *Derek Jarman's Garden*, by the late film director, artist and writer, with evocative photography by Howard Sooley. Jarman sought refuge in his garden, on the shore near Dungeness nuclear power station, a setting with no boundaries, where everything is on an edge: shingle, sea, sun, wind all shifting and changing. He put wild with cultivated, made art out of rubbish and the garden a gallery where nature played the most important part.

I often flick through the exhibition catalogue of loosely painted and vibrant images of the winter garden, all flopping cabbages and snow, or the froth of spring blossom and furrows with new growth, at Sandalstrand, created by early twentieth-century Norwegian painter Nikolai Astrup. He spoke of himself as one of the painters 'most attached to a place and to the soil'. Encircled by bird cherries, birches, lilac trees, horse chestnuts, white butterbur, copious varieties of rhubarb, berry bushes and hogweed, the terraced farm and garden was an oasis in the challenging landscape of western Norway. Astrup made twenty-two paintings, including six interiors and still lifes at Strandalstrand and was well known for his rhubarb wine.

Planning for dahlias.

21 January

I am foraging for summer in the seed and plant catalogues. I make dream lists for fantasy schemes of arches tumbling with old-fashioned sweet peas in faded chintz colours; fulsome cottage borders enclosed within an ancient walled garden of mellow red brick, and brimming with foxgloves for romance, and white nicotiana for heavenly scents through starry, starry summer nights.

I edit, re-edit, cross most of it out and settle for three good things which would be marvellous enough to realize: Sarah Raven's 'Emory Paul', a flamingo-pink and frilly dahlia; Chiltern Seeds' *agretti*, thin, leafy green saltwort for dressing with olive oil and lemon ('One head will fill a salad bowl'); and 'the new Italian delicacy you simply must try', known as *roscano* or *barba di frate* in Italy. . . . But then there's Mr Fothergill's fulsome praise of runner bean 'Lady Di': 'A vigorous, heavy cropper of long, smooth and slender beans which remain stringless even when quite mature'.

A gouache-and- pastel painting of my kitchen table, inspired by William Scott.

SEMENTI
Sweet Pea
Peacock
Silver-white buds
Lilac veined blooms
Sow Jan-Apr/Sep-Oct
Flowers May-Aug

28 January
Philosopher/writer Alain de Botton notes in a newspaper piece that it's a pity for his writing timetable that the news has been quite so interesting: 'Takes a long time and a lot of browsing on *The Guardian* website until the pain of achieving nothing trumps the fear of doing something badly.'

At life drawing there is limited time for procrastination but this seems to make my creative paralysis more acute, and so I try to take the advice of Gillian, our art teacher and 'dive-in'. It is about observing the human body in front of me rather than making assumptions and drawing what I think I see. It helps to look at the spaces in between the form to work out shape and proportion. It's about having a stab, working on and into the paper, editing and reforming – just like working with words.

At least in the garden there is little angst involved in picking up a spade and getting stuck in; and, however banal the task ,there's always a sense of achievement.

30 January
Cauliflower, pomegranate and pistachio salad:
Introduced to it by my cooking daughters, I've tweaked this Ottolenghi recipe from his book *Simple,* to include more pomegranate seeds and a handful of torn rocket leaves. It's our current favourite for crunch and colour and good with roast chicken (breast stuffed with bay leaves and butter for rich flavour).

This shallow cream coloured oval dish from a junk shop is a favourite.

Ingredients
l large cauliflower
1 medium onion, roughly sliced
6 tablespoons olive oil
25 g/1 oz parsley plus a small handful each of mint, tarragon, and rocket leaves, all roughly chopped
100 g /4 oz pomegranate seeds
25 g/1 oz pistachio kernels, toasted and chopped
1 teaspoon cumin
2 tablespoons lemon juice
salt and black pepper.
Method
Roughly grate a third of the cauliflower.
Break the remaining cauliflower into florets and add to a separate bowl with the leaves and onion. Add 3 tablespoons olive oil, and spread over a baking tray. Roast for about 20 minutes, until golden brown and cooked through. Leave to cool, then add to a large bowl together with the rest of the olive oil, and the grated cauliflower.

Annabel's garlic, in gouache and pastel – more everyday inspiration from William Scott.

Garden Tools

Many local hardware shops happily continue to be a source of simple utilitarian tools. Mine yields a new wooden broom with rigid bristles made from coconut fibres; excellent for sweeping up today's rose prunings.

This broom lives in the shed, alongside my trowel, spade, shovel, fork, rake, hoe and shears. But secateurs and pruning snips sit in a bucket by the back door so I can nip out and cut flowers and trimmings when I want them. When things need replacing I look out for secondhand vintage shapes such as hardwood and iron trowels, forks and spades, which tend to be better made and cheaper then modern versions.

Mr Campbell's old galvanized watering cans are so much more tactile than the plastic sort, but the above hardware store stocks some very good and functional galvanized shapes if necessary.

On gardening gloves: I prefer the green textured latex variety that have a good grip and do for most garden tasks, even if they're not thorn-proof. For thorny work I have a pair of heavy-duty canvas gauntlets that come up well above my wrist.

For lighter tasks Mrs C. W. Earle, 1891 (see *The Virago Book of Women Gardeners*) has some practical advice: 'Old dog-skin or old kid gloves are better for weeding than the so-called gardening gloves, and for many purposes the wash-leather housemaid's glove, sold at any village shop, is invaluable'. Another bit of useful vintage to look out for on my travels.

On footwear: every gardener knows the earthy-coloured portrait of *Miss Jekyll's Gardening Boots* by Sir William Nicholson, 1920. Years old, well worn, brown men's lace-ups in stout leather with the uppers coming away, they seem to communicate her practical hands-on approach – as if owner, boot and the ground were all of a piece. I am content with black wellington boots and, when the weather is more clement, an old pair of trainers.

I think that weathered and worn galvanized textures look natural and at home in most garden settings.

An oil painting of a pot of narcissi in an enamel bowl (a junk shop find).

10 February

'It was -5°C when we left Watford,' says the delivery driver, as he helpfully scrapes the ice off my front windscreen before craning a huge bag of 'black gold' (compost) into the front garden (big loads are much more economical and can be shared with friends) at 8 a.m. on a freezing February morning.

I trundle a full barrow down the side passage and spread the rich black earth over the first of sixteen beds. Only fifteen more to go. Nourishment too for the side beds, where my mum's peony, together with the pots of agapanthus, will be enriched. Made up of decayed organic matter, it will return slow-releasing essential plant nutrients to the soil and help lighten the sticky London clay.

Apart from the mood-boosting actions of digging and being outside, the pleasing smell of damp earth is also invigorating. Known in scientific terms as geosmin, it is released through the activity of soil bacteria called actinomycetes.

Getting down in the dirt with other bacteria in the soil may also be helpful for our health. Neuroscientist Christopher Lowry has found that the *Mycobacterium vaccae* found in soil can boost serotonin levels in the brain. His work ties in with the idea of 'old friends', a theory that humans co-evolved with useful micro-organisms but are losing those ties in our increasingly urban environments. One of his studies showed that children raised in the countryside among animals and bacteria-laden dust, grow up to have more stress-resilient immune systems and may be at lower risk of mental illness than city kids. (Never again will I complain about the dog's muddy paws messing up the freshly mopped floor.)

Hard to imagine that in a few months the beds will be brimming with plant life . . .

Sunset in Winter: an oil painting of the view from my home studio.

14 February

The English weather is like an unreliable lover: the gloom of the past few days has dissipated in sunlight and a sheet of blue sky . . . what tomorrow will bring is anyone's guess. On mild days like this I hear in the garden the persistent tea-cher tea-cher of the great tits, three pairs usually, in super flashes of blue white and yellow all bobbing and darting around the overhanging silver birch. *Tweet of the Day* on Radio 4 reveals that they have between them around forty vocalizations, and a single bird up to eight. So prolific, but why? One idea is the Beau Geste hypothesis. Just as the hero of the eponymous novel propped up dead soldiers around the fort to make the enemy think he was better defended than he really was, so a single great tit with a large repertoire may convince rivals that there is a whole gang of them to mess with if they threaten his patch.

Glass jars and dried alliums: see-through textures for a natural ethereal look.

16 February
Snow flurries at breakfast. The cold is piercing and I have no great urge to go outside.

I edit more Barbados pictures from my recent trip to the small island birthplace of my grandfather, who was brought up with grassy cricket 'pastures' edged with tamarind trees, and the limpid Caribbean in year-round brilliant shades of turquoise and cobalt. It was an embracing environment of tropical heat, humidity and the trade winds blowing sweet breezes.

A paradise for the small boy playing toy maraccas with shak-shak pods on the wooden verandah of St Clair, who later became a brilliant classics student awarded a Barbados scholarship to Oxford.

I can picture the cold contrast of stone cloisters but can't begin to imagine the icy traditions and the challenge of being the single 'coloured' student of his college in 1926.

I have ordered *Black and British* by David Olusoga and *Black Oxford* by Pamela Roberts for my research into the identity of my father's father – only recently discovered, with great joy and pride.

Recycled green glass jars: favourites at all times of the year.

As I was saying earlier . . . snow is somewhat of a rarity (though far from unknown) in London nowadays.

20 February
Scorchingly cold, but the garden is infused with warming light and soon white frost tributaries on the window panes will melt in a downstream direction. I pruned pink and frothy 'Saint Swithun' rose a week or so ago.

The budding forth of early shoots is not compromised by the frost. I think it's because the garden is relatively sheltered in its inner city suburb cloak of insulation, but even so we experience harsher extremes than the more lowland gardens on the way to the river Thames.

On garden texture, I think my favourite galvanized metal is also the perfect neutral backdrop for a rose arch, and indeed support for any climber. In turn the arch is a very simple way to create the sense of a garden journey, both by focusing the eye through it and by physically experiencing the motion of entering and leaving one part of a garden for another.

New spring shoots on 'Saint Swithun' rose.

Narcissi and hyacinths, an oil painting.

25 February

At least twelve years ago, Ilse gave me a Christmas gift net of narcissi bulbs for the garden. Pleasingly and without fail, lush green clumps push up from midwinter through the sleeping ground under the apple tree, and in a matter of weeks the tight buds unfurl to reveal all the cadmium and lemon yellows of spring. I could be feeling down in the dumps about something, but the returning blaze of life and colour is always of great comfort. (By the way, I am having good results with white scented 'Thalia' narcissi which have recently joined the happy throng.)

Winifred Nicholson's way with simple and brilliant still lifes of flowers sends me to my local garden centre for a pot of forced yellow narcissi, and then I do a painting inspired by her *Cyclamen and Primula*, a simple, luminous study of plants wrapped in tissue paper on the sunny windowsill of the house overlooking Lake Lausanne where she lived with her husband, fellow painter Ben Nicholson.

Ilse's narcissi in bud under the apple tree.

Spring

The clocks have changed, the evenings are growing longer, and all the buds and shoots and green things in the garden are stirring in the new spring season.

'Rems Favourite' tulip – and my favourite, too.

Spring Fever

'In the spring, at the end of the day, you should smell like dirt.'
Margaret Atwood, *Bluebeard's Egg*

The white spikes of scented apple blossom are heady and seductive and reflect Christina Rossetti's words that 'spring is when life's alive in everything'. A flower's sole purpose is to procreate, enticing the birds and the bees and other insects to help – and, much like the bees, we get a buzz from the flowers too.

Monet wrote that he might perhaps have owed becoming a painter to flowers. Their beauty calms and revitalizes us at the same time. The simple geometries in nature are captivating in the beauty of a flower's form, displaying proportion balance and harmony – which we respond to much as we do to rhythm and harmony in music and mathematics.

Scent gives a signal that a flower is ready to be fertilized, extra-important for night pollinators such as moths that follow fragrance trails in the dark. Scents of the garden are entwined in our deepest olfactory memories 'madeleines are everywhere in the garden and surely Proust is its guardian spirit,' reflects Michael Pollan in *Second Nature*. The sweet and voluptuous, almost sickly, smells of bluebells, hyacinths and narcissi propel me back to my childhood and Easter Sunday: I'm wearing a new printed floral cotton skirt, the table is laid with a white cloth, and in the centre is a large cut glass bowl with said blooms just picked from Grandma's garden. There's a chocolate silver-foil-wrapped egg in a box on my breakfast plate and the sun pours through the window, bathing us in warmth and renewing light.

The greens of nature, and particularly the vivid and fresh new greens of spring, are compelling. One theory is that our ancestors may have developed a greater sensitivity to various shades of green, as opposed to any other colour, because they needed to understand all the plants in a predominantly green landscape. It seems we are inherently drawn to the natural greens associated with plant life and exposure to green spaces makes us feel well. Research has shown that looking at green foliage increases concentration and in one study workers asked to lift green and black boxes that weighed the same thought the green ones felt lighter.

'It was one of those March days when the sun shines hot and the wind blows cold when it is summer in the light and winter in the shade.'
Charles Dickens, *Great Expectations*

As green calms us, it's good to bring it inside. I used a dark olive green ('Hopsack' it was called), for the wood-panelled dining room in the early eighteenth-century house we restored in Spitalfields, east London. Somehow the age of the house and its oasis-like setting among urban sprawl and dereliction, called for a rich, timeless and comforting green, that spoke of woodland and nature. At the other end of the scale, acidic lime -greens like the fresh new flowers on the Norway maple tree outside might look like a Hockney painting when set against brilliant blue sky, but this kind of green would be too powerful to live with in large swathes. It's better used as bright shots of colour in a room on, say, cushions or featuring as a solitary tulip on a glossy green stem.

Here's another thing to consider: our affliation with plants and flowers has been primed since our hunter-gatherer beginnings, when survival depended in large part on the dopamine in the brain that gives us the drive to get up and go find this or that berry. In modern life we now forage and hunt among the online wilds of social media, on Instagram and Facebook, programmed by the same primitive reward system, now manipulated for marketing purposes to keep us ever seeking, never satisfied. Turning off the technology and reconnecting with nature in the smallest way can be a very good thing to help us feel more grounded.

I can't resist little pots of grape hyacinths (muscari) with their cobalt bells arranged on lime-green stalks like clusters of upside-down grapes. I think every home should have one, or two, or three . . . so little outlay for so much visual pleasure.

18 March

The grass is lush and shiny, ready for the first trim of the season. Despite being a child of the suburbs, I belong to the laid-back school of lawn care, unlike some lawn-obsessive friends who hyperventilate over every stray dandelion and any yellow patch caused by a visiting dog. (Hands up, it was mine.)

Gardeners' Question Time helpfully suggests a daily dose of tomato juice neutralizes the smell of dog urine. I enjoy the more shaggy look to my grass, without being quite as vehement as garden writer Michael Pollan, who finds the weekly mow 'laborious and futile' and describes a lawn as being 'nature under culture's boot'.

Margaret Renkl, in *The New York Times*, also makes the case for neglecting lawns, on the scientific basis that when scorched by weed- and moss-killers lawns are drained of their bioversity. I'm sure she's right.

Spring sunlight on 'Thalia' narcissi, perfect in terms of simple beauty, colour and scent. I like to cut one or two stems for the table.

Spring hyacinths, depicted in pastels.

21 March
On the radio I listen to 'The Trees', a poem by Philip Larkin written to celebrate the spring.

> The trees are coming into leaf
> Like something almost being said;
> The recent buds relax and spread,
> Their greenness is a kind of grief.
>
> Is it that they are born again
> And we grow old? No, they die too,
> Their yearly trick of looking new
> Is written down in rings of grain.
>
> Yet still the unresting castles thresh
> In full-grown thickness every May.
> Last year is dead, they seem to say,
> Begin afresh, afresh, afresh.

Larkin sees life and death in spring, for trees and humans alike. I am mindful of the apple tree, battleworn after surviving its collapse a few years ago, hollow in parts of the trunk, with wizened and tattered branches – yet here it is again, with sprigs of green buds.

New spring details: hazelstick wigwams and gravel paths with a permeable lining which halts the weeds.
Tulip 'Purple Dream' is one of the first to bloom.

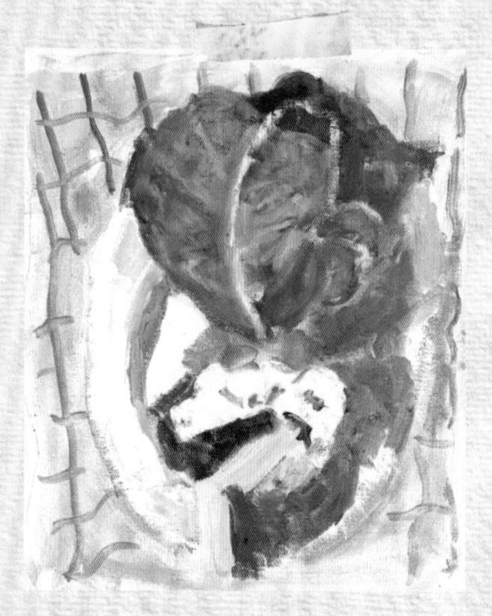

Spring greens from a trip to the veg stall, an oil painting.

29 March

From my desk by the window I have been marking daily the steady greening of the old horse chestnut and how its floppy green leaves are unfurling like chicken's feet. There will be no conkers in the autumn as its days are sadly numbered because the insurance company declares that the roots are causing my house to subside. It does seem a terrible waste of a tree. In consolation the blue suburban skies are brilliant today, and as I read a book on the grass, all shiny and velvet lime-green in sunshine at lunchtime, a passing bee almost tickles my ear and a languid vapour trail overhead dissolves into a meringue froth, all fluffy and delicious, like the burst of cherry blossom next door.

I'm making a pea and mint soup with a handful of fresh spring greens (or spinach leaves) thrown in for more nourishment and spring vigour. (By the way, spring greens, cabbages and so on, are, of course, usefully available all year.)

Pea and mint soup:

Chop one onion, four garlic cloves, two peeled potatoes and add to a pan with l litre/1¾ pints vegetable stock. Bring to the boil and simmer for about 15 minutes, until the potatoes are soft. Add a handful of chopped fresh mint leaves, and a handful of spinach leaves or spring greens; 500g/1 lb frozen peas (petits pois are sweeter), with the juice of half a lemon, and bring back to simmer for five minutes. Season with salt and pepper. Whizz to a creamy texture in a processor or with an electric blender. Delicious with a dollop of sour cream, crème fraîche or yoghurt.

Horse chestnut leaves unfurling in a vintage enamel jug.

The August
For George

5 April
Almost overnight the apple tree has exploded in a froth of pinky white blossom. If this is intoxicating to me, think how overwhelmingly desirable it must be to the bees and other insects who are on their primal missions to pollinate and help procreate. We will be in blossom heaven for about ten days if storm free and then as the petals fall the fruit will set.

I have always liked to believe that 'Mr Campbell's apples' might be a rare heritage variety. Last autumn I sent off 'three ripe, not over-ripe, fruit specimens and a shoot with representative foliage' to the National Fruit Collection at Brogdale. I was very excited to learn that our tree is the 'Keswick Codlin'. According to the Brogdale report, this variety was: 'Found on a rubbish heap at Gleaston Castle in Lancashire. Introduced by nurseryman J. Sander of Keswick; known 1793. Cooking to juicy, creamy froth or purée; hardly needs any sugar . . . also makes a refreshing, juicy eating apple. One of most popular early cookers in the nineteenth century; grown also for market around London and Kent up to 1930s, but now only found in gardens. With profuse early blossom and neat habit, trees are highly decorative; in 1890s recommended for arbours, tunnels, etc.

I'd say Mr Campbell's 'Keswick Codlin' does exactly what it says on the tin.

PS Van Gogh's 1888 blossom paintings of the orchards around Arles are glorious canvases of limpid blue spring skies grass greens and ethereal blossom whites. Saw some at the Royal Academy exhibition years ago, but think a trip to the Van Gogh Museum in Amsterdam for a closer look is long overdue.

The froth of apple blossom beguiles us and the bees alike.

I enjoy spring indoors with cut blossoms from the apple tree.

Tulip Stories

Along with pragmatism and sensible bicycles, tulips are very Dutch. Well, the former maybe, but the tulip originated in Central Asia and was the most prized flower of the Ottomans (think of tulip motifs on fabrics, carpets,Iznik tiles and miniatures). Its name, *tulipa*, is derived from the Turkish *tulbent* (turban), which clearly the flower resembles. The popularity of the tulip in Islamic art is also connected to its shape, as *Allah* written in Arabic script resembles the tulip form.

Botanist Carolus Clusius brought the tulip to Europe in the mid-sixteenth century. Already endowed with exotic roots in Ottoman culture, the tulip took off as an even more highly valued commodity when it became infected with a virus that altered the colour pigments in the cells of the petals, to create extraordinary patterns of flames and feathers. These mutated 'broken' tulips fuelled the speculative obsession of 'tulipomania'. In 1636, at the peak of tulipomania, one bulb of a 'Semper Augustus' tulip could fetch the price of a fancy Porsche today. It all crashed, as things do, a year later in 1637. 'Semper Augustus', decorated in red and white striations, and its near-rival 'Viceroy', white streaked with purple, both feature in Ambrosius Bosschaert the Elder's *Flowers in a Glass Vase* of 1614.

The tulip remains probably Holland's most popular flower – go see the spreading blankets of colour in fields packed with thousands of tulips at the Dutch Tulip festival of Keukenhof, from March to late May each year. But broken tulips won't be among them as, ironically, it's not economically viable for commercial growers to have infected tulips. (By the way, Rembrandt tulips are technically imposters, being genetically bred.) A few broken favourites exist, and may be found at some tulip shows – including the yearly Wakefield and North of England Tulip Society's show in May of English florists' tulips.

In the old-fashioned meaning of the word, a florist is essentially a floral enthusiast. The Wakefield society has been going since 1836 and is a remnant of the florists' societies that working men belonged to from the mid-eighteenth century on, during the industrial revolution in northern England.

Getting up close with the tulips on a sunny spring afternoon – an oil painting.

They were a response, perhaps, to grinding work and migration to the cities which had weakened traditional social ties and pursuits. The societies were devoted to a range of flowers including tulips, auriculas, carnations, pinks, polyanthus and pansies. Between 1750 and 1850 just about every noteworthy town in the north of England had its tulip society. With imbibing high on the agenda, early florists' societies were based in pubs and taverns: the Halifax growers met at the Shoulder of Mutton Inn and those from Leeds congregated at the Golden Cock Inn at Kirkgate for example. Members of the Wakfield society continue to grow 'broken' tulips, but now show in the more sober confines of the Normanton CommunityCentre. However, brown beer bottles continue to be the exhibit vessel: after all, nothing must detract from the detailed scrutiny of the blooms. 'You don't plant them for mass, you plant them for close-up observation, because of their beauty,' says one enthusiast. Judging rules are rigorous.

There are three colour groups: 'Bybloemens', white base, mauve or purple petals; 'Rosens', white base, pink or red petals; 'Bizarres', yellow base, yellow petals marked with red, brown or black. And three types of tulip: 'Breeders' are the original tulips, unaffected by the breaking tulip virus and in solid white or yellow. 'Flamed' and 'Feathered' tulips have been 'broken' by the virus and are marked with appropriately described flamed or feathered lines of contrating colour. Flowers will be marked down if there are 'skips' where the colour hasn't taken, or if the markings are too heavy.

As for army recruits being inspected on parade, there are more hoops: petals must not flip out at the top, or bend in to close off the top of the cup. The base colour of the flower must be 'cleanly' white or yellow, on top of which the darker colours are laid, the stamens boldly black and the petals rich and strong. (I can feel an idea for a 'Tulip Off' show, with behind the scenes angst in the tulip beds.)

It's worth noting that the Hortus Bulborum in Limmen, northern Holland, with thousands of heirloom bulbs, is the only museum garden where you see tulips dating from the 1590s, including 'Duc van Tol Red and Yellow', the oldest known cultivar, together with seventeenth-century daffodils. I must combine this with a Van Gogh excursion. Finally, do read Deborah Moggach's novel *Tulip Fever* (and catch the film, too).

Tulips in Bermondsey

On the subject of tulips and the greening of dull cityscapes, go see the cheering spring blooms from the autumn planting of ten thousand tulip bulbs, given by a grower who couldn't deliver during the Covid pandemic, in the old city docks area of Bermondsey and Rotherhithe in east London, where volunteers from the Greening of Bermondsey are carrying on the legacy of Ada Salter, a redoubtable woman who in 1920 was elected mayor of the then borough of Bermondsey – the first woman mayor ever to be elected in London. Her Beautification Committee planted thousands of trees and ten thousand tulip bulbs, as well as building public baths, children's playgrounds and better housing. This became known as the 'Bermondsey Revolution'.

The Observer wrote in 1931 that: 'Outside the Royal Parks it would be difficult to find anywhere with such masses of colour; when the tulips and daffodils are over they will be followed by armies of dahlias, geraniums and antirrhinums.' The *Evening Standard* described Bermondsey as 'the most optimistic place in London, because of the flowers' and pointed out that Bermondsey factory girls wore flowers pinned to their coats.

10 April – teatime

When the sun was out an hour ago the tulips were a mass of flapping fluttering petals, as if each flower had unfolded like an upturned umbrella bathing in the sunshine. It clouded over and the petals closed like sunbathers reaching for a blanket to wrap up in.

Pinks and purples are my tulip colour themes. I have considered two- tone shows of purple and white or dark pink and orange – like stand-out party frocks at the races – but I prefer the ease on the eye of a more overall colour look. And of course, the glossy greens of spring are such a good match with pinks and purples. I once went through a phase of all-white flopping Parrot tulips, but more as a styling concept in my white country cabin fantasy.

I like the spare look of a few tulip stems in the green ribbed vase on the mantelpiece. You appreciate each one so much more than if a whole bunch is all rammed in together. I am reminded of the – now quite iconic boldly coloured 'Tulpaner' cotton print with a repeat of simply drawn tulips and leaves by mid-twentieth-century Austrian architect and designer Josef Frank, whose 'Paradise Prints' ('gorgeously coloured birds and butterflies fly over materials covered in flowers and berries, edible fruits on large trees, as if in the garden of Eden') were shown at the Museum of Textiles in Bermondsey a few years ago. I was pleased too to see Frank's orginal watercolour for the tulip print alongside.

My all-purpose green junk vase put to good use with tulips from the garden.

Tulips, forget-me-nots, and an allium reaching for the sky.

Tiptoe Through the Tulips

Advice is to plant the bulbs pointy end up in late autumn (I've waited until midwinter with no adverse effects), 10–15 cm/4–6 inches deep and about 7 cm/ 3 inches apart. Tulips repeat flower but this is a hit-or-miss affair depending on weather conditions. Best to dig up after flowering, put in a shady spot and let the foliage die back. Then store the bulbs in a dry, dark place until ready to plant in autumn. I have to say that, in a belt-and-braces way, I also order new bulbs. I'm not bothered with early tulips and go for mid- or late-season tulips.

Mid-season tulips

'Blue Heron': Fringed, 60 cm /24 inches. I love the violet-purple colour and feathery pale violet crystalline edging of this tulip

'Blue Parrot': 50 cm/ 20 inches. More fuchsia-pink than blue, with extravagantly 'fimbriated' ragged petals. Best for cutting, as they flop in the beds.

'Flaming Flag': Triumph, 55 cm/22 inches. Rembrandt-style, white with purple flames.

'Jackpot': Triumph, 45 cm/18 inches. Deep maroon-purple with feathered white edging.

'Negrita': Triumph, 45 cm/18 inches. Vivid purple blooms on hardy green stems.

'Purple Dream': Lily flowering, 55 cm/22 inches. More dark pink than purple. The slender globlet profile unfolds in the sunshine to an elegant flopping star -like shape with a white centre.

'Rems Favourite': Triumph, 45 cm/18 inches. Rembrandt-style. My favourite of favourites: ivory-white with beetroot- purple feathering like a gorgeous raspberry ripple ice cream.

'Spring Green': Viridiflora, 50 cm/ 20 inches. White with cup-shaped blooms with prominent green 'flames' and more subtle white detail. Good in pots.

Single late tulips

'Bleu Aimable': 60 cm/24 inches. Lavender-mauve flowers that pale as flowering proceeds into late spring.

'Recreado': 50–70 cm/20–27 inches. Rich dark purple petals. A good contrast for lighter pink tulips – or team with the even darker 'Queen of Night'.

'Violet Beauty': 55–75 cm/22–30 inches. Goblet shaped lavender flowers with delicate pink veins and tall stem. Like 'Bleu Amiable', flowers well into late spring.

FROM LEFT TO RIGHT: 'Rems Favourite' (in foreground); watercolour of tulips; 'Violet Beauty'; 'Jackpot'; 'Bleu Aimable'; tulip drawing; 'Recreado'; 'Purple Dream'; an accidentally 'broken' 'Blue Heron' bloom (in foreground).

Anatomy of a 'Rems Favourite' Tulip Flower

Beautiful petals dying back and a lesson in flower reproduction. The six stamens (the male parts) consist of a filament and anther at the top which produce the pollen. This is transferred to the pistil (the female parts) at the centre which consists of the stigma at the top which catches pollen grains that travel down the style to the ovary at the base. If compatible the pollen will fuse with the ovules inside the ovary to form a fruit; the ovules become the seeds of the fruit.

The gentle decline over a few days of a 'Rems Favourite' bloom: the petals and stamens fall away to reveal the pistil and seed pod of the flower.

Wild and Weedy

'What is a weed? A plant whose virtues have not yet been discovered.'
Ralph Waldo Emerson

I share the garden with the plants that please me; many serious gardeners would be amused by the way in which I don't discourage so-called weeds unless they become particularly rampant. I am kind to the weeds – after all they are simply wild plants growing in the wrong place – but that is a very subjective view, I suppose, not only of the weeds, but also of me. Consider the sapphire-blue forget-me-nots that run riot through the tulips and alliums, giving the garden an ethereal country garden quality. I love their unbounded growth – and, if I think they're getting above themselves, I pull the plants before they go to seed and again before they get going in the early spring. Or there are the floppy cabbagey leaves of the comfrey plants that increase in number each year but make a pleasing backdrop to the beds in summer. They are satisfying to paint, too. And, again, I edit them out with a fork when needed. The dandelions, which, like many garden weeds, are so good to the bees and the butterflies, are allowed to shine in sunny crowds, then make their fluffy clocks. Some so-called weeds, such as enchanter's nightshade, with its small green leaves and flower spikes of white flowers and pink buds make their claim in a subtle and pretty way. The trailing green sticky garlands of cleavers (or sticky willie, or goosegrass) that get so friendly with the long grass under the apple tree have a delicate, rambling appeal. But even this tolerant gardener has a pet hate, and that it is the dreaded creeping buttercup, low-growing with quite pretty yellow flowers and leaves. I loathe the way it digs its roots into the cobbles around the herb beds

When we moved in it was spring, and I remember that the bright clumps of bluebells helped give the tangled and neglected garden a sense of possibility. 'A paradigm of art – this grace of blue conjured out of the moil of roots and rotting leaves and mite-stirred soil.' (Gerard Manley Hopkins)

I like to imagine that 'Mr Campbell's' bluebells are remnants of the wild ones that carpeted the ancient Great North Wood, of which this garden would have been part long before developers arrived with their constructions.

Each spring vivid stabs of violet bells with flopping leaves push up under the apple tree. I pick them and put them on the table for some welcome spring colour. The English bluebell used to be known as the jacinth or harebell and, in folklore, as granfer griggles, goosey ganders, crowtoes, and fairy cups. Elizabethan herbalist John Gerard thought the flowers had 'a strong sweet smell, somewhat stuffing the head'. The pear-shaped bulbs produced starch to stiffen Elizabethan ruffs and the flower stalks produced a glue for binding books or fixing feathers on an arrow.

Judging by the fact that we can catch only a slight whiff of sweetness, I think what we have here in the garden are hybrids, crossed with the Spanish bluebell, a scentless invader (albeit established in Briain for at least the last three hundred years), that woodland conservationists fear is driving out the native flower. I learn that English bluebells have a distinct sweet scent, narrow leaves, deep blue (sometimes white, rarely pink) narrow tube-like flowers, with the very tips curled back, and flowers mainly on one side of the stem only, drooping or nodding at the top; the anthers with the pollen are usually cream. Spanish bluebells have broad leaves often as much as 3cm wide. Growing from all sides of the stem, the conical or bell shaped flowers with spread out tips are paler blue (or white or pink) and have an almost imperceptible scent.

It is good to know that there are various supplies of sustainably produced English bluebell bulbs, and because they are a protected species the brilliant blue haze of a bluebell wood in spring should not become a thing of the past any time soon. West Woods in Wiltshire, Hayley Wood in Cambridgeshire, and the woodlands along the Ribble valley east of Preston in Lancashire, are all sites well worth visiting.

Biodegradable pots for planting seedlings directly in place.

22 April

More foraging from the house belonging to Wilson (my dog Rosie's best dog friend) for heavenly-scented English country garden sprays of lilac.

In my worn cloth-covered charity shop find copy of Constance Spry's *Simple Flowers: A Millionaire for a Few Pence*, lilac, not unreasonably, is one of her few pence flowers. 'It grows in the less favoured positions in country gardens as well as in many a dusty town yard. Lilac massed in a box or a bowl, set low on coffee table or stool, is not only good to look down on but for such an arrangement short-stemmed pieces are suitable, and these last better than longer branches; neither of course do well if one neglects to remove the leaves from the flowering stems, not of course discarding useful sprays but arranging them among the flowering heads though detached from them.' Timeless advice from 1957.

23 April

Planting up pots and trays with runner bean, dwarf bean, rainbow chard, rocket and basil seeds. The shed suffices as my greenhouse – well more of a frost-free space than a space with all-round light, but on sunny days I bring the seedlings outside to help them on their way. I have in mind (another fantasy) a proper nineteenth-century-style greenhouse, light-filled, with pretty window panes, gabled ends, and greenhouse benching. A place to commune with the trailing tomatoes and scented geraniums, and something good on the radio.

Lilac from Wilson's garden.

25 April
It must have something to do with all the purple tulips in the garden that the 'what shall we eat' neurons alight on smoked mackerel with beetroot on toast, for a really easy tapa and early drink down by the shed where the sun hangs out until the very end of the day. The plate combo of purple on blue echoes the tulips amongst the blue throng of forget-me-nots. The garden inspires in so many ways!

A sprig of alkanet, a handful of lime-green lemon balm, so sherbety in smell, and a rhubarb leaf are all that's needed for a on the spot wild garden posy to decorate the mantelpiece or table .

It's a good time to make comfrey fertilizer. I turn to Griselda Kerr's rich nuggets of gardening advice in her book *The Apprehensive Gardener*: Remove flowers regularly to keep the foliage fresh. Pick the leaves, throwing away tough stalks, and put in a bucket to make a rich potash feed for most plants. Weigh down the leaves with a brick, topping up with fresh water adding 10 litres of water to every kilo of leaves, stir occasionally and leave until it has broken down into a liquid (which stinks). Strain, bottle, keep cool and dark. Dilute again 10.1 to use; it should be the colour of tea. Put used leaves on the compost heap.

For the smoked mackerel with beetroot:
For 4 people mash 2 fillets of smoked mackerel with 2 tablespoons yoghurt, juice of half a lemon, and a handful of chopped parsley. Season with salt and pepper.

Spread on toast smeared with a little olive oil and top with slices of cooked beetroot. For more bite and heat, add a tablespoon of grated fresh horseradish (or horseradish sauce) to the mix.

The wild bunch: alkanet, forget-me-nots and Mr Campbell's bluebells.

Kitchen window view on the spring garden – an oil painting.

29 April

Tulips are in peak colour with cobalt forget-me- nots and bluebells blazing against a luscious green on green leafy backdrop. And snug in their protective sawn in half plastic water bottle surrounds the dahlia tubers can sprout new leaves with less risk of snail incursion.

It's the sort of day when a friend who doesn't garden or one who thinks they can't might come round and complement me on my green fingers or green thumb as its known in the US. Opinionated garden Eleanor Perenyi has this to say: 'People who blame their failures on not having green thumbs, usually haven't done their homework. There is no such thing as a green thumb. Gardening is a vocation like any other – a calling if you like. But not a gift from heaven. One acquires the necessary skills and knowledge to do it successfully (or one doesn't)'. Personally, I think there's more to it than expertise – it's an attitude, an acceptance of. and understanding of, nature's whims.

And it's through trial and error, and simply having a go, that one finds a balance, a middle way between pushing nature too far and giving her too much ground – not unlike bringing up children, I suppose. The main thing is not to aim for perfection: too much is out of our control, on both fronts.

Spring is when the garden's colours are at their most vivid.

Alliums

An allium-themed watercolour.

I like to know my onions, and particularly the ornanamental alliums which bring height and simple architectural beauty to the late spring garden. Alliums, from the *Amaryllidaceae* family, include hundreds of species, among them the cultivated onion, garlic, scallion, shallot, leek and chives. I like to experiment with different ones, but all have the magical quality of appearing as a star-shaped burst of green some time in midwinter (if there hasn't been fierce cold). The leaves fill and some time in early spring a central bud appears and shoots up on a slender green stem to continue its journey skywards. That in itself is mesmerizing to watch over the weeks, but there's more to come with the gradual unpeeling of the flower bud at the top of the stem, now maybe a metre/39 inches in height, and the emergence of a glorious pompon ball of star-shaped or dangling bell-shaped flowers.

Plant the bulbs 15 cm/6 inches deep and 20 cm/8 inches apart to prevent the flowers being too crowded. Avoid excessive soil moisture and split and divide large clumps in autumn or spring.

An allium bud unfurling bit by bit until it is a round, fluffy ball packed with hundreds of tiny purple star-shaped florets.

Allium cristophii (star of Persia): stellar beauty in both flowering and dried forms.

24 May

The bees hug the allium pompon heads, vying for space with the butterflies. My alliumscape features different heights and textures.

This year *Allium cristophii* (star of Persia), native to Iran and Turkey, has self-seeded and spread into even more gorgeous vast and spherical pale lavender flower heads – like a burst of exploding fireworks. At a height of around 30 cm/12 inches, the lacy matrix of star-shaped flowers (20 cm/8 inches in diameter) knits seamlessly with the lavender and nigella. The faded flower heads continue to look spectacular through summer, until they collapase from fatigue. I save a few for bringing inside.

At around 80 cm/30 inches, *Allium hollandicum* 'Purple Sensation' has small purple heads of star-shaped flowers and is a useful visual layer beneath the tallest allium megastars (fewer of these, partly because of expense and partly because of the less-is-more-powerful strategy).

Allium 'Gladiator' fits its name, with spectacular giant 15-cm/6-inch pinky-purple pompons on stems 1 metre/39 inches high; I can also recommend 'Globemaster', again with large deep violet pompon flower heads; and 'Ambassador', one of the tallest architectural alliums at 1.5 metres/5 feet, with 20-cm/8-inch pompons densely packed with hundreds of small star-shaped purple flowers.

Then there's *Nectaroscordum siculum,* Sicilian honey garlic, pretty and dangly, sending up several stems 1.2 metres/4 feet tall with dangling bells of pale lilac flowers for the bees to do their acrobatics on. There's more interest when its spent flowers close and move skyward, resembling a sheaf of arrows.

No shrinking violet: *Allium* 'Ambassador'.

In sky and sea inspired watery blues and greens, this retro junk shop cotton seersucker tablecloth makes regular appearances on the table, and for spreading out a picnic feast on the ground.

29 May

Humid and all-enveloping warmth meant I could sleep with the window pushed up wide all night. It rained hard and I woke to a garden all sodden and drooping.

Now in the evening still the blackbird is pouring out its flute-like song, always so cheering, and beady-eyed robin flits to the rose arch. The grass smells sweet and the sky is fading to indigo in the east.

Over a few days the bright pinks of the tulp blooms have faded, and they combine with the pompon alliums to give the garden a more overall look of mauves and purples on a sea of green. With the thermometer at 23° C/74° F, it feels very much like summer is here. The garden is our default room outside: a place for repose or for laying a simple table.

Everything's going on in the late spring garden: mauves and purples of alliums (flowering chives, too); bursts of blue nigella; roses 'John Clare', 'Saint Swithun', 'Winchester Cathedral', and 'The Ancient Mariner'; and the silvery spikes of the cardoon.

Summer

The garden is voluptuous and running wild after the last few days of monsoonal weather. 'Constance Spry' roses, with cabbagey whorls of pink, scented petals, flop heavy with raindrops, and the robin is dotting about the soggy brick path, in and out of the tangle, looking for worms.

Summer luxuries: the first blowsy and scented 'Constance Spry' roses.

Here Comes Summer

Summer is as much a a state of mind as a season – I say. All year round I organize and decorate my home with light, bright, optimistic colours, fresh white floors and airy texture.

Then – summer arrives in the garden, with the first roses and the washing dried and warm from the line!

The luminous and fresh vivid greens of spring have subsided, making way for the more multi-layered rich greens of summer fields, lawns and grassy hedgerows. Around the middle of the season the greens flatten, but splashes of colour remain, and towards the end there's beauty in the drying out, bleaching and yellowing of seed heads and plants as they decay.

The garden space is a canvas for colours and ideas as much as the rooms are inside and my palette is, of course, an edited one. I try to be quite ruthless therefore in not allowing colours that jar with my vision of things – while, at the same time trying to be practical and resourceful within a budget.

I was amused to read an Instagram post by decorator Abigail Ahern, queen of chocolate walls and moody evergreen gardens. 'Darn those pink hydrangeas. Hate pink in the garden, bought them having been told they were green,' she says. I completely sympathize, in so far as I would feel very glum in an all-green shrub garden, or a garden with my most-hated combination of scarlet, blue and orange (think municipal park planting pre the influence of king of perennial planting Piet Oudolf).

My go-to soft green colour, which I match from a chart at the paint shop, works both inside and out for walls, fences and pieces of furniture. It's like a kind of undercoat to set the scene, knitting together the garden and indoors.

LEFT: As a man of practical and simple tastes, Mr Campbell, the previous owner of the house, chose (and unwittingly and very pleasingly, bequeathed to me) various beautiful and useful household items – a shed full of old lawnmowers, wartime kindling he bagged up and stored in the cellar, an array of watering cans and this pair of simple wooden cupboards, still lined with newspaper sheets c.1957 and now the perfect vehicle for my flower and plant displays.
OVERLEAF: The early June garden is brimming and luxuriant.

3 June
Arrived back from Olhão last night to find that in contrast to baked terracotta, sardine scents and whitewashed walls, the garden has taken off in an almighty verdant and lush growing phase.

At breakfast white hailstone peas ricochet off the roses and alliums; all is humid, drenched and soaking – it's like being in a Caribbean 'passover' storm without the steaming heat. My arms are braced against the thorns as I prop up the sodden rose branches, all flopping and drooping, with more wire; there's the fresh smell of mint and I feel the warm tickle of aromatic lavender spikes against my bare leg.

Coco cat lies in one of her snug, dry grassy hides by the fence, sneakily waiting to pounce on a bird. Our robin, perched on the beansticks, is having none of it and goes into the agitated 'tsee tseee' alarm call.

Lots of daily grind stuff – bill-paying, washing, a difficult conversation, and a trip to vet for expensive dog treatment. After all this my reward will be to pick some floral girls for the table: 'Gertrude Jekyll' and 'Constance Spry', with their hugely scented pink rosettes, work so well with a few stems of nigella, whose lime-green feathery leaves top the purply-blue flower like the ring-setting for a glamorous jewel. The pink/blue/lime-green combination is a favourite, and I think the delicate ribs of the glass jar (a vintage French jam jar from a local antique shop – not expensive) make a good counterpoint for the exuberant bunch of blooms.

Pink, blue and green: a simple glass jam jar stuffed with 'Gertrude Jekyll' roses and sprigs of nigella.

Jewel-like nigella stems and leggy alliums, two important garden components creating both vertical structure and delicate detail, all through the summer season and beyond.

Summery greens in the garden, an oil painting.

4 June
I have just bought a new pair of snips for deadheading the roses, to encourage more blooms and to be able to pass on the task to my youngest daughter, Gracie, the current garden help. You should deadhead repeat-flowering shrub roses and once-flowering shrub roses which don't produce hips. Do not deadhead hip-producing roses if you want hips in the autumn/winter.

The David Austin rose catalogue supplies useful advice: 'Pinch or cut off the finished flower, just below where the base of the flower joins the stem. Leave any remaining buds or blooms to continue flowering. Do this as required throughout the flowering season.'

We gather the fallen petals and dry in a sieve to make scented potpourri for the house.

'Constance Spry' roses: a mass of pink, scented blooms – to be enjoyed fleetingly, as she flowers only once in the year.

Garden Roses

The roses recommended here are very little trouble as long as they are pruned heavily and well nourished: minimal fuss for unbounded visual joy and scented pleasure.

'Constance Spry': Climbing Shrub rose, old rose hybrid, strong spicy fragrance, large cabbagey pink blooms; flowers once.

'Gertrude Jekyll': Climbing Shrub rose, old rose hybrid. Strong fragrance. One of the first to bloom in the garden, and repeat-flowering.

'Grace': English Shrub rose, apricot blooms, medium tea rose scent; repeat-flowering (opposite).

'John Clare': hardy medium Shrub rose, light fragrance, deep rose pink flowers; repeat-flowering (above right).

R. filipes 'Kiftsgate': Rambler, masses of fragrant white flowers; flowers once. I don't deadhead immediately because clusters of delicate rose hips look great with other garden pickings.

'Madame Alfred Carrière': Noisette rose, creamy white blooms, sweet fruity fragrance; repeat-flowering. I have had on north-facing walls here and in Spitalfields.

'Saint Swithun': English Climbing rose, soft pink blooms, medium-strong fragrance; grows around the rose arch beautifully; repeat-flowering (above left).

'Teasing Georgia': English Climbing Shrub rose, creamy yellow rosette-shaped flowers with a sweet fragrance; repeat-flowering (above centre).

'The Ancient Mariner': English Shrub rose, large many-petalled pink blooms, mid-pink, paling toward the edges, medium-strong myrrh fragrance; repeat-flowering. (Name inspired, of course, by Samuel Taylor Coleridge's *The Rime of the Ancient Mariner* – although not sure what the two have in common!)

'Winchester Cathedral': English Shrub rose, white flowers. delicate fragrance; repeat-flowering.

A watercolour of garden rose pickings.

10 June

I grow a luscious fuchsia-pink peony which is a wonderful memory of my mum, who grew the original plant. When she died in 1999 I divided it and brought this bit home here. The only thing is that it is a flash in the pan – reavealing its frilly, flopping delights for only a week or so. The same goes for the double white one that a friend gave me fifteen years ago (nameless also until I was writing this book, when plant guru Tony Lord identified it, from the photograph opposite, as *Paeonia lactiflora* 'Festiva Maxima'). I don't mind cutting this for the table because the wet weather ruins the blooms so quickly, even if they do look moody and photogenic on the way to their demise. At least inside they have the chance of a longer life. If you are wary of buying those shop peonies that look so promising with their about-to-open buds, but which often don't go on to open fully, here are some tips from my years of styling and coaxing flowers as accessories for photographic shoots:

Trim to the length you want. Remove excess leaves and turn upside down in warm water for half an hour. Then pull back just the leaves and the bud should gently plump (like one of those comforters that are air-packed, and when you let a little of the air out, the entire comforter expands), and you can go on pulling the petals back, one by one.

Paeonia lactiflora 'Festiva Maxima', a glorious peony grown by Margaret in the borders of her meticulous south London garden (our dog, Rosie, put me to shame peeing on and burning up the lawn when we stayed with Margaret one summer while a shoot took over our house for the Jack Dee comedy series *Lead Balloon*). The tubers she gave me produce a frothy show without fail each summer. The blooms, white with a tinge of pink, in a battered metal jug, make a pleasing juxtaposition of soft textures and simple utilitarian form.

Pale and Interesting

When I want to create a white summery look for the table, or want the mood in the room to be more understated, I'm grateful for the tendrils of white jasmine and clusters of 'Félicité Perpétue' Rambler rose which tumble accidentally over the garden fence from next door. There are also beautiful pickings of my own Rambler 'Kiftsgate', with huge sprays of white single flowers, and the ever-reliable Climbing 'Iceberg' whose bridal blooms go on right into winter if there's no frost.

I don't usually go for elaborate vases and containers. A spare, simple and utilitarian look tends to be my choice: glass jam jars, rounded pudding basins, enamelled jugs – that sort of thing. I might be a huge Constance Spry fan on account of the eponymous rose, but I surprised myself when, on a visit to an out-of-town fair, I chose a cream pottery boat-shaped 1950s Constance-Spry-style vase (complete with wire stem holders). Almost out of my comfort zone but not too fussy, and perfect, it turns out, for the simple white display here. For those who don't know it, I would recommend the terrific biography *The Surprising Life of Constance Spry: From Social Reformer to Society Florist*, by Sue Shephard. Spry dominated British floristry from the late 1920s until her death in 1960, creating displays that were sophisticated and innovative. She might heap together masses of a single flower, she might offer up a huge, lavish arrangement of artful and layered blooms. She took pleasure in using ordinary plants: daring (for then) concoctions of curly-leaved kale, huge stems of rhubarb and silvery fronds of globe artichoke. She shunned 'proper' vases, instead raiding her clients' cupboards for suitable vessels, repurposing everything from meat platters to baking tins.

Quite early in her career, frustrated with being unable to buy the sort of vases she wanted, Spry designed her own, the best known of which were made by Fulham Pottery. They were often matt-textured, boat-shaped or wide-necked . . . and I think I might be saving up for one.

White on white – Rambler rose 'Félicité Perpétue' in my boat-shaped vase in Constance Spry style is simple and timeless.

11 June
I have borrowed the colour and voluptuouness of early summer – mint, marjoram and a single bloom of 'Constance Spry' – for an aromatic and scented scene in a simple white ceramic jelly mould on the white bedroom mantlepiece.

Outside, the garden's early summer evening beauty, brings to mind more of Michael Pollan's ideas in my embarrassingly well-thumbed edition of *Second Nature*: his view is that gardens are simultaneously real places and representations. Among other things a garden is a passage to somewhere else – to the personal and shared past its scents evoke, to the distant places to which its forms allude.

Roses, alliums, and lavender about to burst forth in purply blue scented spears.

12 June – foxglove fantasies

Sunken lanes brimming with wild roses and banks of purple spikes leaning into the windscreen, shady woods where shafts of sunlight sweep through masses of slender foxglove bells (aka witch's gloves, bloody fingers, gloves of our lady, dead man's bells, or fairy thimbles) are summer visions as much as sweet cottage gardens with plantings at the back of borders for height and colourful structure.

I had an idea for an array of purple and white foxglove spikes among the alliums, but, as is the perogative of the gardener, didn't quite get round to it. This potted common foxglove (*Digitalis purpurea*), a spotted and speckled bee magnet from the local garden centre, is no less of a pleasure and has bloomed for the last two weeks. It's coming to an end but is no less interesting as the seed pods form and the structure takes on even more triffid-like shape. An eerie plant indeed, it is lethal at very low doses: 'The active principle in digitalis may destroy life and leave no appreciable sign,' says Dr Gerard in Agatha Christie's *Appointment with Death*. The bells, however, made excellent goblets for our childhood dolls' tea parties, or fairy hats for our fingers. (Not sure how we're still here: we were disregarding our own doctor father, who had schooled us in the foxgloves' deadliness along with that of the medical supplies in the hallway chest that we were forbidden to touch.)

Have a look at Van Gogh's portrait of his physician Dr Gachet, who is holding a foxglove stem. Apparently he prescribed digitalis (a heart drug) for the artist's epilepsy, and the story goes that it affected Van Gogh's eyesight and accounted for his yellow period of painting.

A pot or two of foxgloves from the garden centre – instant summer colour.

19 June – a welcome invader

Peony poppies (*Papaver somniferum* Paeoniiflorum Group) have blown in on the wind, the wing, or perhaps even from someone's discarded poppyseed roll . . . but what luck that they are in my kind of pink! the frilly whirl of petals is so like a peony in form, but with a very much more interesting set of grey- green frilled leaves. And, yes, they're related to the opium poppy. The seeds are edible but the rest of the plant is poisonous. These poppies are annuals, but the flowers can be just as large as those of the perennial oriental poppy, and they bloom later than their perennial cousins. Deadhead peony poppies to get re-bloom. Follow the stem of the spent bloom down to the first set of leaves and cut just above.

A drawing of peony poppies.

20 June – almost midsummer

The garden is all about looking forward, so here I am, enjoying the pale blue twilight still persisting at 9.30 p.m. with a recently delivered bulb catalogue: I am planning my autumn deliveries of spring bulbs for next year. As well as the tulips, alliums and narcissi, I might consider irises; I have a vision of a pale blue swathe at the bottom of the garden where it's more wild and tangly. Dutch iris 'Silvery Beauty' with deep blue and white colouring and bright yellow blotch would look pretty and is good for cutting. But my heart really lies with the splendid Tall Bearded irises, grown from rhizomes, which are taller, more flamboyant and definitely more paintable. I might go for pale powdery blue 'Jane Phillips'. It will be an experiment, as Bearded irises prefer full sun and sharp drainage, and though I have the aspect, the clay soil might well be a challenge too far.

Self-seeding poppies add to the free-flowering, shaggy and serendipitous nature of my garden.

Dreaming Gardens

My elemental need to grow things, to be the mother, as it were, of seeds that I nurture to make food to eat, is fulfilled by simply filling some trays and compost each year and watching the progress from seed to the moment when I shake the earth off freshly picked leaves for lunch. Beatrix Potter's delicate watercolours of plump cabbages vegetable garden lusciousness in the *Tale of Peter Rabbit* fed my childhood imagination, together with a miniature garden set with which I spent absorbed hours planting tiny plastic cabbages and leeks in neat rows of plastic beds, separated by cardboard pieces of brick or crazy paving . . . ah the benefit of play . . . a more worthwhile investment in the future than passing through exam hoops, perhaps? And a play that I did realize on a grown-up scale in Andalucia, twenty years ago, when my children were small, and we shared a big vegetable garden with Camilo,who knew all about the soil, a man of it, who created the requisite orderly rows of tomatoes (the blueprint for growing tomatoes worth eating) from seed handed down each year; chickpeas, aubergines, cucumbers, peppers and beans. Crouching down in the cool shade in the early evening, among the herby-scented avenues, picking fat tomatoes with stretched-smooth pink skins or stubby rough-skinned cucumbers was all I needed for peace and calm after a long day of human mothering. Searching for local salad and vegetable supplies is built into the itinerary of all my travels. It is bliss to buy farm produce from stands like the Pumpkin Patch under a rough wooden pergola on the North Fork of Long Island. On a summer visit to the island of Barra, a four-hour voyage on the CalMac line from Oban – where ten days of Caribbean-blue sea and bleached white beaches belied the fact the weather is dreach most of the time and the latitude too far north for a plentiful growing season – we were delighted to discover the polytunnels at North Bay Garden, stuffed with all kinds of salads and herbs. We brought back a big paper bag filled with oak lettuce and rocket to eat with an island crab linguine at our rented house. Closer to home, I bike to the Brockwell Park community greenhouses, for half an hour of heaven wandering along old red brick walls heavy with espaliered pears, past shady bowers, wild flowers, at least four different varieties of gooseberry, and raised beds of plumping vegetables. Volunteers man a stall with a clutter of plants, herbs and local honey.

Some of the best advice for growing comes from Charles Dowding, the 'no dig' organic gardener, who simply mulches soil with compost annually rather than laboriously digging it. I have always 'dug' like this, more from laziness than for any other motive, but I'm pleased to be on Dowding's side. His solution to costly raised beds is to cover the patch earmarked for your veg with cardboard and then cover that with a compost mulch 5 cm/2 inches thick. He says you can grow most veg successfully that year although you may have to wait a while before you can grow roots, such as carrots and parsnips, giving time for the cardboard and the soil underneath it to break down. Consider too, the rise of the rooftop farmer. Singapore is a hymn to concrete and metal but look closely and you can see farms mushrooming across the city scape: on the roofs of malls and carparks, in schools, warehouses and even the site of a former prison. On the fourth floor floor of an office building, Mr Benjamin Swan grows plumes of kale and lettuce under LED lights, planted in a substrate with not a speck of soil in sight.

23 June
I have planted out runner bean seedlings at each of the four hazel-stick wigwams at the corners of the flower and herb patch, tying string around the frame at intervals to train the shoots. In the side bed the peastick structures fashioned from the twiggy tops of hazel branches were supposed to be home to sweet pea seedlings, but the order was cancelled by the grower. Instead I've planted dwarf beans 'Tendergreen', but, again, not all is working as planned, because already the snail and slug brigade have been out on the slide. I'm luring the beasts with beer traps, and hope yet to reap the benefits.Then for a recipe, adapted from Margaret Costa's *Four Seasons Cookery Book*.

Ingredients

450 g/1 lb beans, steamed and drained
250 ml/1 cup *soured cream*
Salt, pepper, nutmeg
40 g/1½ oz soft breadcrumbs tossed in butter

Method

Toss the beans in the soured cream. Season. Turn into a buttered dish and top with the breadcrumbs. Bake at 190°C/375°F for about 15 minutes, until golden and bubbling.

'French Breakfast' radishes with a pot of sea salt – an oil painting.

25 June
Swiss chard is filling out well – but slugs have destroyed the 'Little Gem' lettuces. However, I have been seeding in succession and hope that the next batch will not succumb: the weather might be less humid. I'm having a go for the first time with a radish colour mix of 'Purple Plum', 'Scarlet Globe', 'Sparkler 3', 'White Turnip' and 'Zlata'. But I will still find space for pink 'French Breakfast', so pretty on a blue plate with a pot of sea salt to dip it in.

Dwarf beans 'Tendergreen' wind around peasticks – and there's the promise of crunchy 'French Breakfast' radishes to dip in sea salt.

Be Friendly to the Bees

'Bees are the batteries of orchards, gardens, guard them.'
Carol Ann Duffy

I try to make the garden friendly to bees, butterflies and other insects, with wild flowers and flowers with accommodating shapes for nectar-gathering. Together with its spilling over and shaggy appearance, this makes the garden an exciting proposition for nature in general, including the birds (jays, woodpeckers, magpies, sparrows, goldfinches, wrens, blue tits, wood pigeons) and (the quid pro quo of creating a patch of country in the city), the not-so-welcome urban fox.

As I write on the green-painted potting and all-purpose table by the open door of the garden shed I feel as if I'm being checked out by four or five fluffy bee sentinels with yellow and black stripes hovering at ankle level by the entrance. They've been here over many summers, and I'd always assumed, in that way we don't always look as closely at things as we could, that they are stray honeybees from a neighbouring beekeeper. But the white tails identity then as bumblebees of which, I read in *The Bee Book* by Fergus Chadwick, there may be up to four hundred in the nest below my shed. A wonderful thought, to have such valuable pollinators literally on my doorstep. More curious now, and down on my hands and knees, every minute or so I see a bee duck down into a dark void, like the entrance to an illicit nightclub, just under the floorboard where it meets the edge of the brick path. The one-in-one-out policy continues with a couple more who shortly zoom out for more partying on the purple thistles which are also magnets for all the other bees that visit. I've noted ten or fifteen types this summer – amazingly a fraction of the more than 250 species of wild bee in the UK, which include more than two hundred different solitary bees, twenty-five types of bumblebee and only two types of honeybee. I've toyed with the idea of hosting a beehive . . . as a way of encouraging bee populations and also enjoying my own honey. Urban beekeeping has flourished in recent years,

The globe thistles are the main attraction for local bumblebees.

with many museums, charities and businesses creating colonies on their roofs, but conservationists say that urban beekeeping is putting wild, solitary bees at risk. With colonies of around fifty thousand, honeybee hives are intensively farmed factories of the bee world: 'Keeping honeybees is an extractive activity. It removes pollen and nectar from the environment, which are natural resources needed by many wild species of bee and other pollinators,' says González-Varo from Cambridge University's Zoology Department.

Honeybees are active for nine to twelve months and travel up to 10 km/6 miles from their hives, potentially out-competing wild pollinators.

Verbena bonariensis plays host to a visiting honeybee.

It's increasingly important to make sure that our own backyards are rich sources of pollen and nectar for wild bees. All of the following are good for summer: allium, borage, buddleia, campanula, catmint, comfrey, cornflower, echinacea, foxglove, globe thistle, hollyhock, honeysuckle, lavender, poppy, sedum, sweet pea, thyme, verbena.

4 July – butterflies

Summer has passed its vivid prime, but remains rich and beguiling. Shimmering spikes of paling lavender and pompon thistles are humming with bees; a red and black speckle of ladybird nestles in the lime-green of the lemon balm.

I'm on a butterfly hunt (to see, not to capture). Ah – I have spotted a large cabbage white, flapping above. Then a flighty brown one with a black spot on each wing. I chase the flighty brown one. It's a gatekeeper, I later find out in the guide.Delicately fanning its wing on a thistle, like a piece of costume jewellery, is a stunning peacock – with dramatic 'eyes', spot markings, that terrify its predators. Must look at Van Gogh's evocative *Field with Butterflies* in vertical stabs of green summer colour and white paintmarks.

The UK has fifty-nine species of butterfies, fifty-seven residents and two regular migrants: painted lady and clouded yellow. Butterfly numbers have depleted; as gardeners our task is to provide nectar throughout the adult phase of their lives.

Especially good butterfly foods are: buddleia, field scabious, hebe, hemp-agrimony, knapweed, marjoram, sedum, red valerian, verbena, wallflower 'Bowles' Mauve', and wild marjoram.

And don't be too precious about getting ride of the dandelions and brambles. They're essential butterfly nourishment too.

Looking back to the colour and vibrancy of the midsummer garden, an oil painting.

20 July
The sun's out, the air is warm and the garden beckons. The wet has galvanized the rhubarb, all curly and spreading for a second crop. The roses are also having another flowering: 'Gertrude Jekyll', 'John Clare', 'Iceberg' (but of course not dear 'Constance Spry', who has had her one and only flowering of the year).

The apples have swollen in the rains, I hope they don't pull down the tree. Like some flashy blue topknot of a tropical bird (and indeed we do, in London, have our very own feral luminous green and red parakeets), the cardoon thistle heads are the most eye-catching element in the garden at the moment. I have propped up the top-heavy stems with a stout stick and string. Similarly, the gangling purple verbena is tall and visually engaging (and also in need of propping).

LEFT: Architectural cardoons are my favourite way to create height combined with colour and drama in the garden – and they are undemanding, provided they're well watered (and well propped). ABOVE: Cardoons, painted in acrylic.

Summer Herbs and Aromatics

'The tender leaves . . . and sprigs fresh gathered put into wine or other drinks, during the heat of summer give a marvelous quickness. This noble plant yields an incomparable wine, as is it that of cowslip flowers.'
Entry on lemon balm from a seventeenth-century herbal

It goes without saying, that such is my (over-) preoccupation with the look and feel of things, that if the most delectable of herbs were of of a form and colour that offended my eye, then it would not be offered a place in my garden. It is another of nature's boundless joys that that my favourite herbs are as beautiful to look at as they are deliciously edible, scented or useful outside of the garden.

I'm drawn to small but perfectly formed creations, such as Michael Pollan's herb garden, in his book *Second Nature,* surrounded by a wall of roses. 'Almost all of the plants are set out in a simple symmetrical garden pattern, nothing as intricate as an Elizabethan herb garden, but carefully balanced so that one side of the garden offers a mirror image of the other.' He suggests that the symmetry of the garden and the precise balance of its plants seems to still motion and suggest repose. The trouble (or maybe it's not really something to be troubled about at all?), is that there's only so much order and structure I can handle, as much in the garden as in life. So my herb and flower garden is a hybrid of the formal and the more free-flowing: the basic shape of sixteen beds with four standard roses at the centre is a framework for an impressionist paintbox of colours, and textures; a magical and ethereal space enhanced by the interplay of light and shadow.

Basil Aromatic, aniseedy, herby basil is the one herb I cannor resist buying from the shops throughout the year, until I can plant my own seeds outiside in early June and raise a crop for summer salads. This year I have planted up a pot with Ocinum basilicum ' Genovese'. The young purple and green seedlings are pushing up daily from a sheltered and sunny position by the shed. Here's our family recipe, an all-round pesto to pour on to pasta, eat with a tomato salad, or spread on toasted bread:
Combine, in a processor or using a pestle and mortar: 100 g/4 oz pine nuts; a handful of basil, finely chopped; 2 garlic cloves chopped; half a handful of grated Parmesan cheese; 150 ml/10 tablespoons extra virgin olive oil, sea salt and freshly ground pepper.

Bay Growing with incredible stealth since I planted it as a tentative young seedling a decade ago, the rounded smooth bay bush, now as big as a single-man tent, by the kitchen door is a visual triumph of dedicated trimming with the shears. It also benefits from a good cut-back in the spring to bring on fresh new shoots and leaves. On a smaller scale,f or some formality in a small garden or on a balcony terrace, I recommend a topiary bay ball atop a stem – like a giant lollipop. With a fresh, green, herby scent, part oregano and part thyme in character, bay leaves are the underpinning for bouquets garnis to flavour soups, stews and other dishes.

Chives Grass-green clumps of chives, the smaller onion cousins of my towering alliums, prettily fringe the borders with bright purple flowers. Taking kitchen scissors to snip a handful of chives for salad was my task as my mum's kitchen hand, a job taken on for me by my own children.

Dandelions Sometimes known as the arugula of the north, dandelions share with rocket a bitter peppery flavour and serrated leaves, but they are part of a different family, the Asteraceae. The Royal Horticultural Society devotes a detailed page to how to control this 'weed' (a dandelion tap root is deep and perfectly engineered to resist removal) but I welcome their yearly appearance in the garden's potager; the brilliant green serrated leaves are rather like beautiful ornamental lettuces .

The early 1960s was not a great moment for wild foraged food in the south London suburbs and so, Peter Rabbit style, we picked dandelion greens purely for the guinea pigs (although my father would daringly eat raw oysters from the rocks on family holidays to Bordeaux – but that was different: it was France). Wiser, and much older, I add dandelions to salad in the beginning of summer; as they become even more bitter, there's nothing to stop me from rustling up a dandelion pesto. The yellow flowers are also edible, useful decoration for a spring birthday cake.

Dill Along with the spare, tall. dangling beauty of verbena, I think that every flower and herb patch could do with a dill plant, whose lime-green feathery fronds spread skywards culminating in delicate umbels of yellow flowers. I have an old faithful that has reappeared for the last five or six years (it's strictly an annual or biennial, but mine grows like a perennial) at least 2 metres/7 feet tall. Its desiccated lacy form stays on in the garden until winter and becomes an even more fairy-tale structure when dusted with an early frost. I stuff fish with the feathery anise leaves,and tear them into small pieces to decorate smoked salmon or smoked mackerel pâté. And, of course, the seeds can be used in vinegars and teas. You can see some flowering dill decorating my desk in a jug of garden pickings on page 149.

As a timely offshoot, today I came across a piece about asafoetida . . . a spice whose name betrays its most notable characteristics: *asa* being resin in Persian and *foetida* Latin for smelly. It is an umbellifer and its cousins include dill, parsley and chervil – all known for their fresh fragrance and flavours. Asafoetida, by contrast , is rich and earthy (OK, like the egg and raw onion sandwich an annoying person might eat next to you on the bus). However, when heated in fat it becomes something more aromatic and delicious – supplying the suphurous notes in poppadoms or a curry or a dahl.

Lavender Lavender's glorious haze of purple blue-flower spikes in summer is breathtaking (as is the scent delicately perfuming the dog's coat after her morning sashay to pee). And then there are supplies for filling lavender bags after trimming at the end of summer. After which, lavender, as a perennial, goes on to create life and structure in the spare depths of winter.

Lemon balm With its sherbet lemon scent and vivid lime-green leaves from spring onwards, lemon balm is good for trimming into soft topiary-like balls, which work well with the surrounding pinks and purples of spring tulips and, later, the alliums. I like to add sprigs to posies, and to mixed flower and herb table decorations. A few leaves will give a lemony flavour to a salad, and also to iced water amd other summery drinks. (Not to mention that I am very tempted by the medieval custom of strewing lemon balm over the floor to impart a sweet scent underfoot).

Mint Pinching a glossy bright green leaf between my fingers to release its fresh mintiness, I'm transported to childhood Sunday lunches with hot new potatoes steamed with mint from the garden and tossed in butter. (It also features in cooling sweet mint tea memories from one 'Hideous Kinky' style journey in the late seventies, with seven of us quasi-hippy students driving a decommissioned ambulance all the way from Upper Warlingham, Surrey, to Marrakesh, Morocco). Mint is a wild child and spreads ruthlessly but not if one keeps an eye and rips up the intruders immediately. Put a cutting in water – it will sprout roots quickly – and give it to a friend to grow in a pot.

Oregano and golden marjoram Low springy clumps of oregano (*Origanum vulgare*) and golden marjoram (*O. v.* 'Aureum') make brimming punctuations of greens and yellows in the summer months, and after they have died back in winter it's always a joy to watch the new shoots come up again in spring. You can scatter torn shreds of the leaves of both herbs over sliced raw tomatoes, or add them to a tomato sauce in progress.

Rocket Put off, perhaps, by the bitter, peppery flavour, the slugs and snails let my young rocket (arugula) plants alone. Native to southern Europe, rocket was a medieval salad staple in Britain but, like so many worthwhile ingredients, later became a casualty of plain cooking and stodge. I prefer the serrated wild rocket leaves to the smoother broad-leaved rocket but both revive the feeling of sultry teenage Puglian summers, with bitter rocket salads and *vongole* spaghetti under a shady vine with my Italian exchange family. (Of course, smooth-skinned and tanned also added to the general excitement.) If the winter is mild, the rocket in my garden keeps on bolting and flowering and even just a few leaves add some Italian magic and painterly green juxtaposition to red and orange in a raw winter salad of red cabbage, beetroot and carrots.

Sage Silvery-grey sage is another perennial that looks good all year round in the garden . . . and it is the perfect flavouring for fried eggs, and mashed or roast potatoes.

Thyme Silvery-grey sage is another perennial that looks good all year round in the garden . . . and it is the perfect flavouring for fried eggs, and mashed or roast potatoes.

28 July – vintage looks

I'm in a retro mood: vintage glass jug and vase (both junk finds) filled with sprigs of lavender, verbena and thistles . . . evoke a kind of classic and timeless look on a green and purple theme. I think of a favourite Colefax & Fowler chintz with a pretty pattern of climbing geraniums, all soft mauves and greens, that I made into a loose chair cover; the kind of fabric most at home styled in sweeping pelmets and festoons in the drawing rooms of sleepy grand country houses. The effect of classic floral print on a chair in my simple white sitting room, without a festoon or a pelmet in sight, is surprisingly timeless. Similarly, Lady Seton's 1927 advice for a gardening outfit might sound old-fashioned but is entirely modern in its practicality. 'While about the subject of outfit, I think that for a great part of the year, an ideal gardening dress for women, is a short tweed skirt, made very wide, so that one can step across plants without injuring them. A loose jumper made of khaki or brown flannel (for half an hour in wet weather will take the shine off a romantically becoming jumper in pale colours), a gardening apron all pockets, a pair of thick shoes or boots and a light scarf tied over one's hair. Hats are dreadfully in the way and if quite uncovered and unshingled, our hair catches in every twig, like Absolam's.' (*The Virago Book of Women Gardeners*)

I realize, too that my watercolour on the right happens to be in the suffragette colour scheme devised by Emmeline Pethick-Lawrence in 1908: purple for loyalty, white for purity, and green for hope . . .

Another watercolour of garden rose pickings.

Purple and green pickings from the garden: lavender, globe thistles, and verbena.

Living Outside

In summer the garden becomes an outdoor extension of the house, reflecting my ideas for natural textures, and a sense of intimacy and informality. Even if it's just putting a chair in a sunny spot to watch the bees play in the pollen of the frothy white 'Kiftsgate' rose bush, or to seek shade on a bench under the apple tree on a hot afternoon, the pleasure of being outside in the elements is the most joyful thing.

Post-pandemic the joy of outside has never been quite so pertinent, whether it's enjoying a garden, terrace, balcony or like last weekend's treat, spritzers on my friend Jane's front doorstep and all the swifts from around swooping and sweeping above in the twilight. Shiny office towers once crammed with seven thousand workers are seeing falling rents as homeworking becomes the new normal. And designers and architects are inspired both to create new natural light and ventilation solutions for inside and to reorganize public outdoor spaces. For example, Studio Precht has proposed the 'Parc de la Distance', an outdoor space in Vienna that encourages social distancing and short-term solitude.

Equally relevant, I recently re-read Alain de Botton's *The Architecture of Happiness* (prompting ideas on how we live and how we might change things) in which he draws attention to *wabi sabi* , the Japanese aesthetic for which tellingly, no Western language, has a direct equivalent. It identifies the beauty in unpretentious simple, unfinished, transient things. There's *wabi sabi* in raked gravel, rain falling on leaves, rough weathered stones covered in moss and lichen. I guess the way I see the beauty in the everyday resonates with the *wabi sabi* attitude even if it has come from a completely different place and experience.

Practical and beautiful

Make the best of outside living with practical, relaxed and portable elements: folding chairs, and tables, rugs and throws that can be taken inside at the threat of rain are very useful. The other thing is not to be too precious. Rather than going for the spanking brand new and glossy, I incline to the side of the weathered and beaten up, choosing, for example, old zinc baths for plant containers and vintage deckchairs in striped cotton canvas.

Old red bricks, cobbles, and worn flagstones are textural surfaces underfoot. Thyme planted between paving stones smells delicious. Pea shingle gravel is a good inexpensive solution for paths and small seating areas, and you can line it with an impermeable lining to stop weeds. (However, this very relaxed-about-weeds gardener is not averse to the self-seeded alliums and dandelions that push up on the unlined gravel 'sun deck' outside the shed.)

In Portugal I laid local terracotta tiles handmade for generations at a small factory in Santa Catarina in the hills behind Olhão. I love the feeling of bare feet against the sun-baked surface when I'm hanging out the washing or relaxing at the end of the day. We made a table from broad and hefty chestnut floorboards (salvaged from a wood yard), which rest on two trestles.When I sit eating or painting a watercolour, the satisfyingly rough surface and close curving lines of the grain are like engaging in a tactile and visual conversation with what once was a magnificent tree.

With heat comes the need for shade. How about a cool front porch with pots of cacti and climbing roses behind a painted picket fence like those fronting the many delectable Victorian cottages I eyeballed in Melbourne? On another trip to visit family in Barbados, I fell for an old colonial wooden house with an airy sea-blue verandah. They said it was like St Clair, the house where my grandfather grew up. In Cartagena, Colombia, I came across simple but clever purple and green public beach tents, a great blueprint for an open-sided retreat from the sun or rain. I love the idea of a shady pergola entwined with the white scented flowers of *Jasminium officinale* . . . or the even more heavenly-scented *Cestrum nocturnum* (Dama de Noche), my choice if it could survive a London winter .

To keep the sun off I make simple removable awnings in cotton canvas, an old sheet, or whatever comes to hand. I hem the sides and stitch on ties

Outdoor living: from washing flapping on the line in Barra, Scotland and a bench in the sun outside a traditional Swedish farm building, to the white-washed quintal at my home in Portugal and Patience Gray's simple plank table still in use in her garden in Puglia.

for securing to hooks fixed in the wall. In the unreliable English summer an inexpensive fold-up marquee can be a life-saver for a small party, and can be jollied up with garlands of flowers.

Then there's the wild thrill of being outside in the chill of a summer's night but warmed by the glow of a campfire, or – more prosaic but none the less effective – a portable fire pit. This feeling is evoked so brilliantly in the loose and vivid paintmarks of Norwegian painter Nikolai Astrup's P*reparations for the Midsummer Eve Bonfire.*

Simplicity is also the theme for eating outside

The thermos rules on picnics. For the wet bank holiday by the sea I might take hot sausages in one, hot soup in another, and one packed with ice for a make-on-the-spot gin and tonic: tonic in small tins and gin decanted into a small bottle; slices of lemon and cucumber. Or, if the weather is looking more promising, a chilled gazpacho can be substituted for the hot soup. Salt and pepper in separate twists of tin foil are lightweight. Similarly pare down the kit you take outside for lunch around the table. A cotton or linen cloth and napkins, a few sprigs of something from the garden, park, or hedgerow give the simplest meal a sense of occasion and are more sustainable than going down the paper plate and disposable cutlery route. Consider serving the main dish or hot potatoes straight from the pan; one or two big platters or bowls for salads are easier than passing round lots of smaller ones. Fill a large jug with ice water and a few lemon balm or mint leaves. On a windy day anchor a cloth with stones at each corner; put tea lights in lanterns or glass storm jars . . . don't waste time trying to keep naked candles alight unless the air is dead still.

Whether it's a Nordic midsummer's night bonfire or a plank table spread with bowls of pasta for lunch,
I think that for living outside one should keep the ingredients simple.

4 August – rain stops play

Packing up the table, and bringing the whole show inside, is something that I've had much experience of through decades of unreliable English summers. But at least we can continue the garden theme with jam jars and glasses stuffed with flowers and herbs.

Above, preparations for a soggy summer birthday when the roses were still going strong and opposite a simple supper for friends, with dahlias and lemon balm. On the menu: roast cod and tomatoes, with a useful Portuguese-inspired all-purpose garlicky green herb sauce: *120 ml/¼ pint olive oil, 2 cloves garlic, chopped, handful of marjoram, mint, basil or whatever I've got growing, chopped, juice of l lemon, salt, pepper. Whizz in a processor. Store in fridge.*

All I need to give style to a simple al fresco gathering is a crisp white cotton or linen tablecloth and some floral sprigs from the garden.

15 August
Idling on a hot afternoon before tackling some work, and watching in close-up a red admiral butterfly sucking up a nectar cocktail with its probiscus straw, balanced atop a tempting bloom of 'Leila Savanna Rose' dahlia . . .

Cooling off in the Brockwell lido – a watercolour painting.

16 August – summer reading
I'm in the mood with *A Month in the Country* by J.L.Carr a novel about the passions of a demobbed soldier in the aftermath of World War 1 that manages to bring alive a rare unlooked-for few weeks of gentle happiness. 'Sometimes listening to music, I drift back and nothing has changed. The long end of summer. Day after day of warm weather, voices calling as night came on and lighted windows pricked the darkness and at day-break, the murmur of corn and the warm smell of fields ripe for harvest. And being young.'

Centre stage: a frilly pink and white 'Leila Savanna Rose' Decorative dahlia. Fence painted in water-based willow-green paint.

Frills and Flounces

A Native Mexican flower, the dahlia is as flounced and frilled as a *traje de flamenca*. It was introduced into Spanish gardens in 1789 and reaching the London Chelsea Physic garden in 1803. Most modern garden dahlias are derived from a parcel of dahlia seed imported from France in 1815. Dahlias have been a mainstay of cottage gardens and allotments, to pick for the table along with the cabbages and beans. I remember my grandfather, fag in mouth, carefully tying his prized rainbow of spiky blooms to stakes with green hairy string. In some elevated garden circle dahlias were long considered a bit naff. In one of my garden clippings, gardener Monty Don regards them as beyond the 'tasteful pale', admitting later that he was wrong (partly from ignorance and partly from prejudice).

American gardener Elsa Perenyi in *Green Thoughts* thought otherwise: 'It hasn't escaped me that mine is the only WASP garden in town to contain dahlias and not the discreet little singles either. Some are as blowsy as half-dressed Renoir girls and they do shoot up to prodigious heights. But to me they are sumptuous, not vulgar, and I love their colors, their willingness to bloom until the frost kills them, and, yes, their assertiveness.'

Dahlias are multifaceted in shape: Single, Ball, Pompon, Anemone-flowered, Fimbriata, Cactus, Semi-cactus, Decorative, Giant Decorative (flower heads up to 25 cm/10 inches across) . . . and more.

And then there are the colours, flamboyant and tempting as a button box or a sweetie pic'n' mix: bronze, dark red to black, flame, lavender, orange, pink-purple, red, white, yellow, bicolour, variegated.

Fashions come and go and now, with the new way of less structured and fussy gardening, is the dahlia's moment. You can Instagram-drool over Charlie McCormick's Dorset vegetable and flower garden with its crowds of pink, purple and yellow dahlias among wigwams of sweet peas and wild flowers. He wins prizes with them at the local Melplash show. Florist duo Terri Chandler and

Katie Smyth of Worm London tuck luscious and creamy café au lait, dinner-plate-sized dahlias into their brides' bouquets and arrangements.

I love the geometric honeycomb pattern of pompon fuchsia-pink 'Franz Kafka', a good compact size for stuffing in a vase or jar on the dinner table. Fimbriated 'Table Dancer 'has frilly vivid purple petals with pink-white tips . . . fabulous against green schemes. You can get mixed bunches from shops for not too much money, and if a colours jars – red is the one that disturbs me – just take it out.

Planting tubers

I have planted dahlia tubers straight into the ground but they seem to suffer from slugs and snails more than if brought on inside and planted out after the frost.

My dahlia PPE is a large plastic water bottle sawn in half and then sawn again at the bottom to make a cylinder to place over the young plant. The important thing is to stake them as they get taller, and to keep removing the flowers to encourage new ones. It's worth it. They have a long flowering season, from midsummer even to the very end of autumn.

21 August – a beautiful tart
More eating al fresco: fig and frangipani tart for pudding (but any stone fruit will do – plums are particularly good).

Ingredients
Crust: *90 g/3½ oz unsalted butter, softened; 75g/2½ caster sugar; 2 egg yolks; 200 g /7 oz plain flour plus extra for dusting*
Frangipani: *100 g /4 oz butter; 100 g /4 oz caster sugar; 2 teaspoons vanilla extract; 2 eggs; 100g/4 oz plain flour; 100 g/4 oz ground almonds*
Garnish: *8 fresh figs (or plums, stoned – quantity depends on size, so adjust)*
Handful of flaked almonds, toasted in the oven for a few minutes until golden brown
Method
Crust: *cream butter and sugar in a processor, add egg yolks, 1-2 tablepoons ice-cold water and flour and process until the mixture gathers into a ball. Roll out on a lightly floured surface and press into a 25 cm/10 inch flan tin with removable bottom. Refrigerate for 30 minutes. Preheat the oven to 180 C°/350° F.*
Frangipani: *cream the butter with the sugar and vanilla in a processor. Add the eggs, flour and ground almonds and process until smooth. Spoon into the pastry shell. Arrange the fig (or plum) halves cut side up, leaving space between neighbours. Bake for about 50 minutes, until the tart is golden brown and a skewer inserted in the centre comes out clean. Leave to cool and then transfer from the tin to a wire rack. Decorate with the almond flakes. Serve warm or cold, with crème fraîche or ice cream.*

Stems of dahlia and cosmos cut for the table.

Georgie's agapanthus in Olhão.

25 August – a family of agapanthus

It all began on my daughter Georgie's eighth birthday in Andalusia, when great friend Rafael (not only a skilled artist and gardener but the most gorgeous-looking man too) arrived from the other side of the cork tree ridge, with a gift of twelve agapanthus plants from a clump he had recently divided. We potted up, Georgie participating with a trowel, and when the Andalusian house was sold a few years later most of them came with us on the back seat of the car to the house in Olhão.

The London crowd migrated later in my hold luggage, against regulations no doubt, between sketch books and a lump of tuna wrapped in a cool bag, and are happily crammed in two large terracotta pots. (I repot with fresh compost once a year, feed with comfrey fertilizer, keep well watered and divide when they get too bushy.)

Agapanthus are a great idea for height and colour if you are restricted for space. I prefer the purple/mauve colours, but white ones are equally distinctive. Like a loved family heirloom, I hope that soon I can pass them back to Georgie, to whom they were originally bequeathed. By the way, it's interesting to note that the Portuguese relations flower in June, but here in London, mid-August is the earliest for blooming in my garden.

Agapanthus reaching for the sky.

Last of the summer rhubarb.

27 August – rhubarb, rhubarb

Saturday was a grey swamp of a day and so I made my way to bed with Bruce Chatwin's *In Patagonia*, a mythical and lyrical journey populated with unicorns, hot winds rolling over the mesquite, dinosaur bones, turquoise lakes, rocks of lilac, rose-pink and lime green, Butch Cassidy's log cabin, hollyhocks and Welsh tea rooms. And outlier personalities too: 'The tenant of the estancia, Paso Rabollos, was a Canary Islander from Tenerife. He sat in a pink-washed kitchen where a black clock hammered out the hours and his wife indifferently spooned rhubarb jam into her mouth.'

At this point I remembered the pile of rhubarb stems downstairs – the last ones from the garden– and thought cake with rhubarb and ginger (freshly grated) would be a fitting accompaniment to my reading. (It was German colonists who brought rhubarb to Patagonia in the nineteenth century.)

Wet rose petals on the path – like delicate tissue paper confetti after a rainy wedding – are also a thing of simple natural beauty.

The Sense of an Ending

On the early morning tour in my nightie, the grass is soaked with dew that sparkles like glass beads. You can see it in yellowing leaves and low evening light – the summer's coming to an end.

Its energy poured into flowering and seeding, the garden is spent, leached and dried out. Nonetheless, there's beauty in decay: cardoons shrivelled and crinkly like scrunched brown paper; pompon thistle heads sucked dry by the bees in faded dusty mauves – I don't wish to sweep away the remnants of summer for the sake of it, but more for a sense of putting things in order, a bit like getting ready to go back to school at the end of the holidays.

As much as the beginning of summer is light and full of promise and anticipation, the end of summer doesn't simply represent an ending, but also the beginning of a new season of retrenching and bedding down. And looking back at the garden's triumphs (flourishing roses and dahlias) and disasters (beans eaten by slugs), there's a wonderful sense of my hard work and nature's serendipity joining in the creation of a simple and peaceful haven.

Long shadows and parched leaves signal the end of summer, but the garden keeps on giving. I fill a glass with dill, a 'Grace' rose and a dahlia stem, and take it to my desk.

Winifred Nicholson

Autumn

For me, between the delights of still mornings and golden sunsets, autumn is the season of mists, bonfires, bulbs, plans, and pumpkin soup.

As the leaves fall from the Norway Maple and Horse Chestnut trees the autumnal scene in the garden features more golden and burnt orange tones.

'Autumn is a second Spring where every leaf is a flower.'
Albert Camus

Autumn is the season of golden contradictions: bountiful and productive yet decaying and dying back. It has the feeling of resource and generosity juxtaposed with a more balancing sense of mortality. Truman Capote is also a champion of the season: 'Aprils have never meant much to me, autumns seem that season of beginning, spring.' And indeed, while it may be the end of long summer evenings, it is the start of the new school year. I too feel the excitement of new autumnal projects if tempered by the slight dread of shorter days and longer nights.

Let's clear it up here: autumn or fall? In ancient times there were really two seasons: winter and not-winter. Summer comes from the Sanksrit word for 'half', defining its status as the other main part of the year. By the sixteenth century, the concept of spring had become more widely recognized, and that then created its own opposite, 'fall'. Later, in England, the Latinate 'autumn' took over; but by then the English had already crossed the pond, taking the world 'fall' with them.

Frost hangs in the air like the sword of Damocles on early autumn days and some nights are so chilly the next day's early morning swim at the lido is a particular feat of endurance. Some days are freeing and warm too, like a throwback to summer. It's mesmerising to watch sycamore seeds spiral like helicopters. I gather seeds from the dried pods and flower heads of the summer nigella and zinnia shows – like all the seeds of the season these are packed with the plants' secret recipes for new beginnings in spring. On my garden agenda there are dahlias to lift, and my mum's deep pink peonies to divide, together with bulb planting. The roses that have shot skyward this summer need to have their sails trimmed (though the main pruning season isn't until late winter). I have ordered two espaliered apple fruit trees for the front garden, and I will dig a bed with well rotted compost to be ready for their eventual delivery.

'If it is true that one of the greatest pleasures of gardening lies in looking forward, then the planning of next year's beds and borders must be one of the most agreeable occupations in the gardener's calendar. This should make October and November particularly pleasant months, for then we may begin to clear our borders, to cut down those sodden and untidy stalks, to dig up and increase our plants, and to move them to other positions where they will show up to great effect. People who are not gardeners always say that the bare beds of winter are uninteresting; gardeners know better and take even a certain pleasure in the neatness of the newly dug bare brown earth.'
Vita Sackville-West

Autumn is also the season of rot and fungi and the giving back of nutrients to the earth. It seems as if the grass bubbles under the tree as the fallen wormy apples, covered with nodules of yeast, give off the fermented whiff of country cider. Mushrooms spring up overnight in the local park, but you have to get there early (to avoid peeing dogs and other fungi fans).

My daughter arrives with field mushrooms to lightly cook in butter with garlic and parsley. I remember olive baskets brimming with with plump ceps on our mushroom foraging after warm days and light rains in the cork oak woods near our home in the Andalusian hills.

In biologist Merlin Sheldrake's book *Entangled Life: How Fungi Make Our Worlds, Change Our Minds and Shape Our Futures,* he describes fungi as nature's alchemists, playing an essential role in transmuting decay into nutrients and keeping entire woodlands alive. These nutrients are sent in multiple directions over the land via the mycelium: a kind of 'wood-wide web'. And, he writes, they might be ecological saviours too, as they are capable of breaking down plastics, petrochemicals and toxic waste, filtering streams and even absorbing radiation.

Afternoon shadows in the shed, and a homely vase of rose blooms.

11 September

The shadows lengthen as the sun's arc drops, day by day. The early September garden is luscious yet on the decline. The grass is ankle deep (if I were a lawn freak, four or five trims overdue), and wet from early morning dews and rain. It is a soft plush landing for the apples. I have already picked three trays of glossy green fruit; they're whoppers this season, through no clever effforts of my own. I love the serendipity of the garden, especially when it works in my favour.

We were so lucky to inherit not only Mr Campbell's glorious apple tree but also the set of wooden shelves for storing the crop.

A Good Apple

I have always liked the taste of the apples from our ancient tree (somewhat tart, but not so sharp as the classic cooker). But since discovering that the tree is the now-rare 'Keswick Codlin' (see page 67), I appreciate them even more . . .

Apple crumble

It might sound a little peculiar but I do identify with the synaesthete I read about who associates stations on the London Underground with food (peanuts, Piccadilly Circus). In terms of taste and texture a bowl of cold, sloppy stewed apples (my mum's go-to wartime remnant for mid-week childhood puddings) evokes the feeling of a monotonous grey afternoon. On the other hand, apples with a buttery crumble top, hot and oozing with juices, teamed with a scoop of ice cream, become exhilarating.

And, if the proof of the pudding is in the eating, then crumble is also comfortingly solid evidence that the best kind of pudding doesn't rely on frou-frou looks; this is both timeless, and right for our in-need-of-comfort times, and will, I assure you, momentarily silence the table.

I am a fond reader of Felicity Cloake's column 'How to cook the perfect . . .' in *The Guardian*, in which she cooks a selection of tried and tested recipes in search of perfect results, and synthesizes the best elements for her take on what makes the perfect . . . whatever it is that week. In the case of crumble, Nigella Lawson claims that rubbing the butter in by hand makes for a more gratifyingly nubbly crumble; and American writer Mary Norwark has stern views on crumble, which she condemns in its English incarnation as dull and insipid (surely that's reserved for the stewed apple?); the English should take a leaf out of American books, she says, by using fresh rather than stewed fruit (something I do anyway). I particularly like Felicity's version because she includes ground almonds in the crumble.

Apple crumble

Ingredients

100 g/4 oz plain flour
50 g/2 oz ground almonds
125 g / 4½ oz unsalted butter, chilled and cut into cubes
35 g/1¼ oz demerara sugar and same amount caster sugar, plus extra for the fruit
9 apples, cored and peeled and cut into chunks
Preheat the oven to 200 C°/400° F

Method

Combine flour and ground almonds and pulverize in a processor until the mixture resembles very coarse breadcrumbs. Sprinkle with a little water and rake through with a fork until you have a lumpy, crumbly mixture. Freeze for 10 minutes and then spread over the prepared fruit in a greased baking dish (I have a small blue and white enamel pie tin that works well). Bake for about half an hour, until golden and bubbling. Serve hot with ice cream (Salted caramel goes particularly well with this) or yoghurt or crème fraîche.

Quinces and red onion, an oil painting.

More Fruits of Suburbia

I'm stirring a bubbling pan of rose hip jelly (balm for thinking) with hips foraged from Saturday's dog walk. I can reveal that my sources are overhanging branches from local front gardens (not strictly pilfering I hope) together with some of the glowing orange hips on the 'Kiftsgate' rose in the garden (best to wait until the fruits are more tender after the first frosts).

More of a secret, although maybe not such a draw when you read on, is a lone medlar tree in the park brimming with fruit that looks like a cross between a rosehip and a flattened apple. It has the unfortunate yet apt name in French of *cul de chien* or dog's arse, and even more potentially stomach-churning to the uninitiated, you can't pick a medlar and eat it ripe from the tree; it needs to be 'bletted' – a process of ripening in a cool place on sand or sawdust until it's squishy and brown.

Also not a fan, D. H. Lawrence described them as 'wineskins of brown morbidity, autumnal excrementa and of leave taking'. Although tasting like a sugary over-ripe date, and delicious with cheese, as is quince paste (see over), the bletted medlar could do with some rebranding.

Apples, alliums and metal jugs for a green and natural kitchen display.

Measuring up Gilly's quinces.

My neighbour Gilly's quince tree is groaning with golden fruit, so with permission granted I set up the stepladder and twist off the furry and fragrant orbs, using a broom to hook the fruit from the upper branches. The deal is that her quinces are picked, and I have the ingredients for making sweet and aromatic *membrillo* (quince paste), the usual companion in Spain for the salty hard sheep's cheese, *manchego*, but equally delicious with any cheese.

Quince paste

Ingredients

3 kg/6½ lb quinces, equal weight of preserving sugar and juice of 2 lemons

Method

Cut up the quinces (peel, pips, core and all), add to a large, heavy-based pan, cover with water and simmer until tender. Purée the mixture with a handheld blender and weigh it, adding an equal amount of sugar plus the lemon juice (for flavour rather than its setting qualities). Simmer the mixture, stirring constantly until it turns a rich red colour. Line shallow baking trays with greaseproof paper and spread the hot paste about 4 cm/1½ inches thick. Leave to harden in a cool place.

I divide the slabs of paste into 10 x 10 cm/ 4 x 4 inch portions, wrap in more greaseproof paper and store until needed.

Rosehip and quince (or apple) jelly

Ingredients

1 kg /2¼ lb quinces or apples, 450 g/1 lb rose hips, trimmed of their ends, preserving sugar

Method

Cut up the quinces or chop the whole apples in quarters, add to a large, heavy-based pan and cover well with water. Bring to the boil and simmer until tender. Add the rosehips and simmer for another 10 minutes. Let cool.

Rig up a jellybag/piece of muslin in a sieve and add the cooked fruit. Leave it to drip into a bowl overnight. Measure the liquid and for every 575 ml/1 pint, add 400 g/14 oz sugar. Return it to the pan and slowly heat until setting point (105°C/220° F) is reached, or test a sample on a chilled plate . Cool and pour into sterilized jars. Delicious with roast meat, especially lamb.

Foraged rose hips and medlars.

22 September
The autumn equinox: equal hours of day and night before the retreat into shorter days.

It's an Indian summer heatwave with unexpected treasures in the garden: creamy and delicate 'Teasing Georgia' rose blooms; jasmine scent wafting from a stray tendril.

Pickings from the garden – rose blooms and 'Kiftsgate' rose hips. It's blanket season too: a worn Welsh checked one on the small wooden sofa makes it feel more cosy in the evening.

24 September
I married on a similar burnished afternoon: a beginning . . . and then an ending three decades or so later in the same month: a kind of reflection of all the beginnings and endings of autumn in the garden. And life goes on.

The cat tickles my shins and jumps on the bench next to me. The dog is on a sniffing check, nose down along the path across to the dahlias and back to my plate, which she licks with entitlement.

29 September
It's Michaelmas –the medieval feast of St Michael the Archangel, when the end of the harvest was celebrated.

The garden is shrinking back. The 'Grace' rose is also doing an end-of-season last hoorah, with a garland of blowsy pale orangey-apricot blooms on a straggling stem. The dessicating cardoons have brown shrivelled leaves like the talons of some mystical beast and yet there's green sprouting too, but temporarily, before the cold sets in.

Architectural cardoons.

Cyanatype of majoram and dill stems.

2 October – photographic blues

Cyanotypes, one of the earliest forms of photography, were created around 1842 by the British astronomer and chemist John Frederick Herschel. What is more, botanist Anna Atkins (1799–1871) used cyanotype printing to make a visual record of British algae in 1843 the first-ever published book of photography.

It is gratifyingly simple to record your garden specimens as beautiful blue (Prussian blue,to be precise), cyanotype prints. I recently bought online a pack of 'sunprint' paper coated with a light-sensitive solution. I picked lavender and nigella stems from the end-of-summer garden (feathers, shells and household objects will work well too), pulled down the blinds in the kitchen and laid them in a pleasing pattern on the paper pinned to a piece of cardboard with a piece of glass on top to keep them in place (a sheet of acrylic or cling film also works).

When it was ready I took it all outside and exposed it to the sunlight for a few minutes, watching the rather miraculous sight of the uncovered area fading from blue to white. On a cloudy day the process can take up to thirty minutes.

Then I submerged the paper in a shallow tray of water for the even more miraculous sight of the white bits turning blue and the blue 'specimens' to white. To get the deepest blue leave in the water for several minutes.

Cyanatype made in the autumn garden of lavender and nigella stems.

5 October – kitchen details
I have been sorting out the cupboards and reassessing. Just as with my favourite garden tools, my kitchen kit is simple and hands-on: worn wooden spoons, basic and functional pudding basins, heavy- based pans with solid bottoms that will not buckle or burn the supper; stainless steel knives (the best I can afford), a simple metal colander for draining the lettuce.

I still rely to a large extent on a basic hand blender and whisk, but I have splashed out on a food processor for its miraculous way of whizzing up soups, slicing and grating fruit and vegetables for delectable salads, and, of course, cake production.

8 October
With another rhubarb crop flourishing well this damp October, I harvest it and make a cake, together with a creamy fool and a roasted rhubarb tart. At odds, it seems, with one of my Instagram gang, who writes: 'I had always thought one shouldn't pick rhubarb after July but I'm not sure why.'

I consult the RHS and see that one is supposed to let the plant build up reserves; also, it is said that after mid-July the stems contain more and more oxalic acid, which is irritating to the gut and to arthritis sufferers. But I guess, as with so much in gardening, this is a general rule of thumb. (Even more reason to garden without rules.)

Rhubarb and ginger cake, the finished article.

Rhubarb and ginger cake
Ingredients
225 g/8 oz rhubarb chopped into 1-cm/½-inch cubes
150 g/5 oz butter
150 g/5 oz dark brown muscovado sugar
2 large eggs
225 g/1 oz grated ginger
150 g/5 oz self-raising flour
Method
Preheat the oven to 180°C/150°F.
Cream the butter and sugar. Add the eggs and the ginger. Fold in the flour and stir in the rhubarb. Spoon into a greased and floured 20-cm/8-inch cake tin. Bake for about 40 minutes, until a skewer piercing the middle comes out clean.

Essential kitchen tools, a painting in acrylic.

Spider Season

My way through the lavender border in the chill of early morning is barred by the ultimate in temporary housing, the most elegant and structurally ingenious short-term accommodation in glistening glassy fibres, and at its centre the architect, a big fat garden spider. And over by the fence spanning the chasm between a thistle and 'John Clare' rose branch there's another huge orb-shaped web perfectly tensioned like a soaring modern skyscraper to swing with the breeze. Two silver birch leaves helplessly teeter in the grip of the sticky web, but their fate is nothing to that of the trapped fly, which is spider-processed, minced and munched in under a minute. Biomimicry, or the use of technology inspired by nature is where a spider web construction comes in useful: for example for its shock absorption capacity at the Estadio de la Plata football ground in Argentina, where the roof is in the shape of a spider's web as a protection against earth tremors.

I read that my familiar garden companions are orb weaver spiders, which build their circular-shaped webs by laying spirals of silk around radial threads. Spider silk is amazingly lightweight: a strand of silk long enough to encircle the earth would weigh less than 450 g/1 lb. It's also as strong as Kevlar, the material used to make bulletproof vests.These garden spiders vary from a pale yellowy brown to dark brown, but they all have a white cross-shaped group of spots on their abdomen. The arrival of autumn marks the official start of spider mating season, meaning the creatures will be leaving their webs soon in search of a dry place to copulate. After mating, the female builds a silken cocoon in which she lays her eggs. She protects this egg sac until she dies in late autumn. No child-rearing work for spiders then. It's not until late spring that the spiderlings hatch.

In art I'm drawn to Louise Bourgeois's giant *Maman* sculpture of an arachnid, one of the first of many spider sculptures she made, as well as etchings and fabrics stitched with web-like patterns of concentric circles and spirals. Her spiders are maternal figures, representing, among other things, her own mother, who was a weaver.

Garden spiders spin their orb shaped webs in profusion around the garden at this time of year.

A Painterly Autumn

The neighbouring Norway maple is a golden and glamorous autumnal feature.

Some gardeners are disappointed when the floral bursts of summer colour are over. For me the colours in the garden are of interest all year round. I don't agree with the notion that the smudge-browns and earth colours of autumn, and indeed winter have limited visual interest: they also appear in so many brilliant hues and tones. Even the least colour-sensitive would have to admit that there is something intense about the few weeks in autumn when the leaves turn and fall and the countryside, as well as all the gardens, parks and commons in the city, are pop-up shows of brilliant reds, yellows and golds. All further enhanced by the long languid light of still autumn mornings and evenings.

I can understand why autumn became famous in the mid-eighteenth century as 'the painter's season'. In her book *Weatherland: Writers and Artists under English Skies*, Alexandra Harris writes that young men on the Grand Tour timed their travels so as to arrive in Rome, and make their excursion to the temples at Tivoli, when the light was most golden in October. Uvedale Price, late seventeenth-century landscape designer and proponent of the Picturesque, considered that autumn was the best season for landscape painting: 'There is something so very delightful in the real charms of spring . . . that it seems a perversion of our natural feelings when we prefer to all its blooming hopes the first forebodings of the approach of winter.' He concluded that the aesthetic appeal must lie in the 'variety of rich, glowing tints' in the dying leaves, as opposed to the uniform greenery of spring and, simultaneously, the quality of light, which unifies this variety in a harmonious whole.

Autumn from my kitchen window, an oil painting.

I like to set up an easel by the kitchen window, and capture the view of the flaming Norway maple, and horse chestnut at the end of the garden. The three paintings which, to my mind, best reflect the voluptuous warm light and textures of autumn are Bonnard's *Autumn Morning*, a country landscape of golden yellows, greens and purple shadows; Howard Hodgkin's *Autumn* in rich red, brown and green abstract marks; and the blazing foliage of David Hockney's *Woldgate Woods*.

20 October

The early morning garden is ruffled and unkempt after a night of strong winds and rain. Norway maple leaves in brown, gold and red litter the ground like pieces of torn paper in a collage.

On the dog walk tsunamis of leaves tumble across the local park. It's a washing flapping in the wind kind of day. I pull my zip up to the top of the puffa. It's extra-layer season too.

22 October

I trim the lavender that I haven't got round to dealing with yet and tie up the spikes in a bundle with string to hang in the wardrobe for some summery scent. A brooding grey sky interjected with partings of sunshine seems to intensify the colours in the garden.

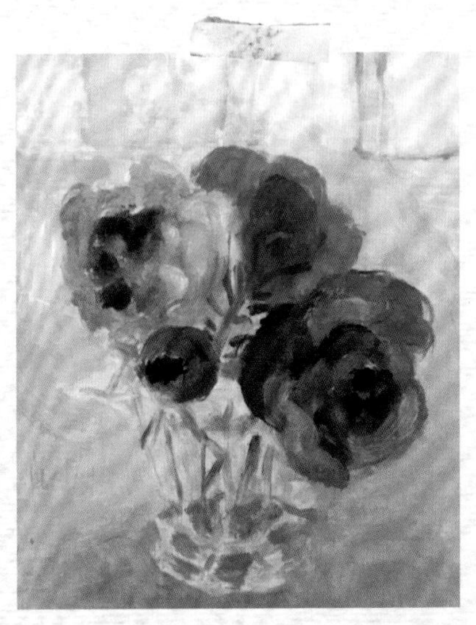

Rosemary, nasturtium leaves, rosehips and roses in a glass from the market, an oil painting.

28 October

I've been away for a few days and have returned to find Jack Frost has blackened the last dahlia. All is cold and damp. Its leaves fallen, the soaring candle flame of next door's silver birch is extinguished, leaving a skinny pyramid of thin twiggy branches. Exciting to see a spotty mistle thrush pecking at fallen apples. At sunset the sky is a vibration of deep blue fading to eau de nil on the horizon.

A soggy 'Eglantine' rose bloom after the first frost.

Alliums for planting.

Planting Bulbs

I set myself quite a task each autumn to fill the sixteen beds of the potager/ parterre with 350-plus tulips and alliums ordered from longstanding favourite companies. As you know, I plant on top of what's already in the ground rather than starting afresh; it's a kind of insurance policy against the strong possibilitiy that many won't reflower, season to season, and indeed many don't.

I know that bulb planting begins in September but I'm nowhere near ready to start then; but equally I don't really want to wait until later than the end of November to put the tulips in (the accepted gardening wisdom is that tulips thrive best when planted in colder soil).The end of October therefore is a suitable compromise, but in turn this goal is often thwarted by a beastly cold that strips me of energy or a spell of freezing weather in which the soil is too hard to work and I have to wait until it thaws. So it's a case of chipping away at the job, bit by bit, bed by bed, over days, even weeks until the last bulb has been bedded (a good 10 cm/4 inches down for tulips) and packed in firmly with the heel of my boot to keep out the squirrels. All very irritating, but so worth it when next spring I'm rewarded with such beauty. It's fitting to learn that the philosopher Nietszche, a keen planter and digger, proposes we look at our emotional difficulties like gardeners – how we deal with the vagaries of gardening can parallel the way in which we 'self-cultivate' our life. 'One can dispose of one's drives like a gardener and, though few know it, cultivate the shoots of anger, pity, curiosity, vanity, as productively and profitably as a beautiful fruit tree on a trellis.'

One bed's worth of tulip bulbs. All mixed up so their random planting produces an equally random array of blooms.

31 October – Halloween

A glowing pumpkin hollowed out and scored with a toothless grin leers from the front window as the ghosts and witches trick-or-treat in little gangs up and down the road. I have some sweets and clementines to offer, rather more modest than the cash that some callers demand! Halloween is actually derived from Samhain, the Celtic festival marked by fires on the day of the autumn full moon: the start of the new year, the death of summer and the onset of winter.

5 November - Bonfire Night

'Remember, remember the fifth of November . . . gunpowder, treason and plot.' In pre-Health and Safety times, the grown-ups appeared unperturbed as we let off Roman candles, bangers and rockets with uncertain trajectories and played around the bonfire flaming with the remains of the guy: an effigy of the Gunpowder Plot conspirator Guy Fawkes (my dad's cast-off trousers and shirt stuffed with paper).

Nowadays – less dramatic but pretty much as atmospheric, in the darkened garden – we stoke a fire pit with logs and swirl fiery arcs with sparklers, eating sausages cooked on the fire and warming pumpkin soup.

I learn from the Bajan cousins that – along with Home Counties comforts like tea and scones, and church on Sundays – the British imposed Bonfire Night on their subjects in Barbados. However, a more local tradition was added: eating *conkies*, cornmeal, coconut, sweet potato and pumpkin steamed in banana leaves. These days, though *conkies* are reserved, more appropriately, perhaps, for celebrating Independence on 30 November.

Pumpkin soup

Ingredients

2 tablespoons olive oil
2 onions, finely chopped
1 kg/2¼ lb pumpkin or squash, peeled, deseeded and chopped into chunks
700 ml/1½ pints vegetable or chicken stock
120 ml/¼ pint double cream
For sage croûtons: *a handful of sage leaves, fried in 2 tablespoons of olive oil until crisp*

Method

Heat oil in a large saucepan, add onions and cook for 10 minutes, stirring occasionally. Add chopped pumpkin or squash and cook for a further 10 minutes, until starting to soften. Add stock, season with salt and pepper, bring to boil and simmer for 10 minutes. Add cream, bring back to boil and then purée with a hand blender.
Serve with hot sage croûtons.

Silver birch leaves: welcome invaders from next door.

November Garden Chat

Once daylight and any hope of gardening has disappeared around tea time, I turn to digging up tips and plant ideas from fellow gardeners . . .

My sister Sarah is mudbound in Somerset, where she has: 'cut everything back and lifted the dahlias to bring into the greenhouse and pot out in spring when the frosts have finished.' She's putting in garlic and onions, adding wood ash from the fire to balance the rich soil, because, last season: 'the garlic was all leaves and small bulbs'.

My stylist friend Nel Haynes' pretty London cottage garden will be packed with treasures next spring. After we have had a good chat about bulbs, she writes:

'I can't locate my actual bulb order but found the catalogue, so I can approve your choice of tulips (!) – and the turned-down edges and scribbles tell me that I'll be expecting some 'Orange Emperor' tulips, plus blue anemones for my shady front garden, snakeshead fritillaries, some VERY pink hardy *Gladiolus communis* subsp. *byzantinus*, a rainbow mix of Dutch irises and finally the wonderful *Nectaroscordum siculum* – previously listed with the alliums, it has a lovely tall stem and lots of bell shaped danglers . . . God only knows where I thought I'd plant that lot in my tiny garden, and it's such a random selection. I was probably a bit tipsy on the computer – very dangerous!'

And over in south London, fashion guru Vanessa de Lisle's keen eye for frock detail is equally honed in her sweet and compact back yard where:

'November is as busy as all the other months. I know its fashionable to leave the garden to die down naturally but I am a manic chopper. I cut down/back, thin out plants, and cut back dead and dying leaves. I need room for the bulbs which I plant new every year. The smaller the garden the more planning is needed to get the balance. I don't like naked gardens in winter so the evergreens are essential. One third is given over to the architecture plants: box balls; holly lollipop trees; choisya, trachelospermum (star jasmine) and even a dull hebe can work in the winter. Camellias look wonderful in pots even when not in flower. Daphne is evergreen and deliciously scented. I have a huge *Jasminum polyanthum* – even if the frost gets the flower buds I love its graceful habit and evergreen leaves. Evergreen ferns are elegant and I also put them in pots. Lavender and rosemary bushes in pots are another winter winner. Anyone can have a midsummer garden but to be pleased with shape and form all year is a huge satisfaction.'

And then it's fun to know what's going on further afield . . . artist Sue Schlabach, who paints the crab apple trees in her Vermont backyard, notes: 'hoar frost from a few days ago . . . today it's full-blown snow out there.'

In contrast, in steamy hot Barbados my cousin, another Su, has spotted a pair of stick insects, in static poses amidst the tropical evergreenery in her garden.

And early summer beckons Down Under, where garden designer Brenton Roberts's cottage garden in the Adelaide Hills is all lime-green euphorbias, purple lavender and grey lamb's ears . . .

I string up lanterns for an alfresco outdoor autumn evening – we wrap up really well and throw a few logs on the fire.

15 November

The plane trees lining the street are stripped of their leaves. Men in fluorescent jackets wield powerful blowers pushing the leaves into papery heaps through which the school kids on their way home take rugby-style kicks and disrupt all the good work.

The science of leaf fall is that as the leaf ages the growth hormone auxin diminishes, and cells at the base of the leaf petiole divide, react with water, then come apart, leaving the petioles hanging by only a few threads of xylem. And then they're airborne on the breeze.

I like to bring autumn inside with a garland of leaves. Cut a length of string and attach each leaf with a simple knot tied around the stem. Pin or tape into place.

Dried cardoons (also at right) and Norway maple seed pods.

18 November

I suppose it's the small girl in me who loves to arrange displays of nature inside. I'm remembering the shelf in my bedroom on which there were shells, fossils, pine cones, and pebbles – my very own natural history museum.

The bowls and plates that I now fill with treasures from the garden, and from my travels (tropical seed pods from Barbados, striped and delicate clams from the tidelines on the beaches near Olhão . . .), as much as being tactile and pleasing to look at, they are also reminders. evoking feelings and memories.

Even if your space is tucked away in a city attic, painting walls and background in softer shades of blues and greens can suggest feelings of sea, sky, and landscape.

Edible and visual treats: sprigs of dill and scarlet nasturtium flowers.

Towards Winter

The spectacular leaf shows have faded and so one notices splashes of more intermittent colour, such as the red breast of the beady-eyed robin which hops over when the chip-chip of my hoe signals that worms are a possibility. Then there's urgent squawking as two pairs of silky black and white magpies rough up an alpha parakeet in the trees. It's surreal to see something so tropical, all lime-green and turquoise flashes, tearing hungrily at the cardoon seed heads.

Scarlet nasturtium flowers are exciting to find among the general greening over by the herb patch on the right, and they look so pretty in a small vase – something to cheer me up on a murky afternoon. There are also marjoram, oregano and the last of the rocket, and, as always, evergreen bay for flavouring dhal for supper. Some mint is hanging on until it is finally taken out by the next heavy frost.

A fading thistle head.

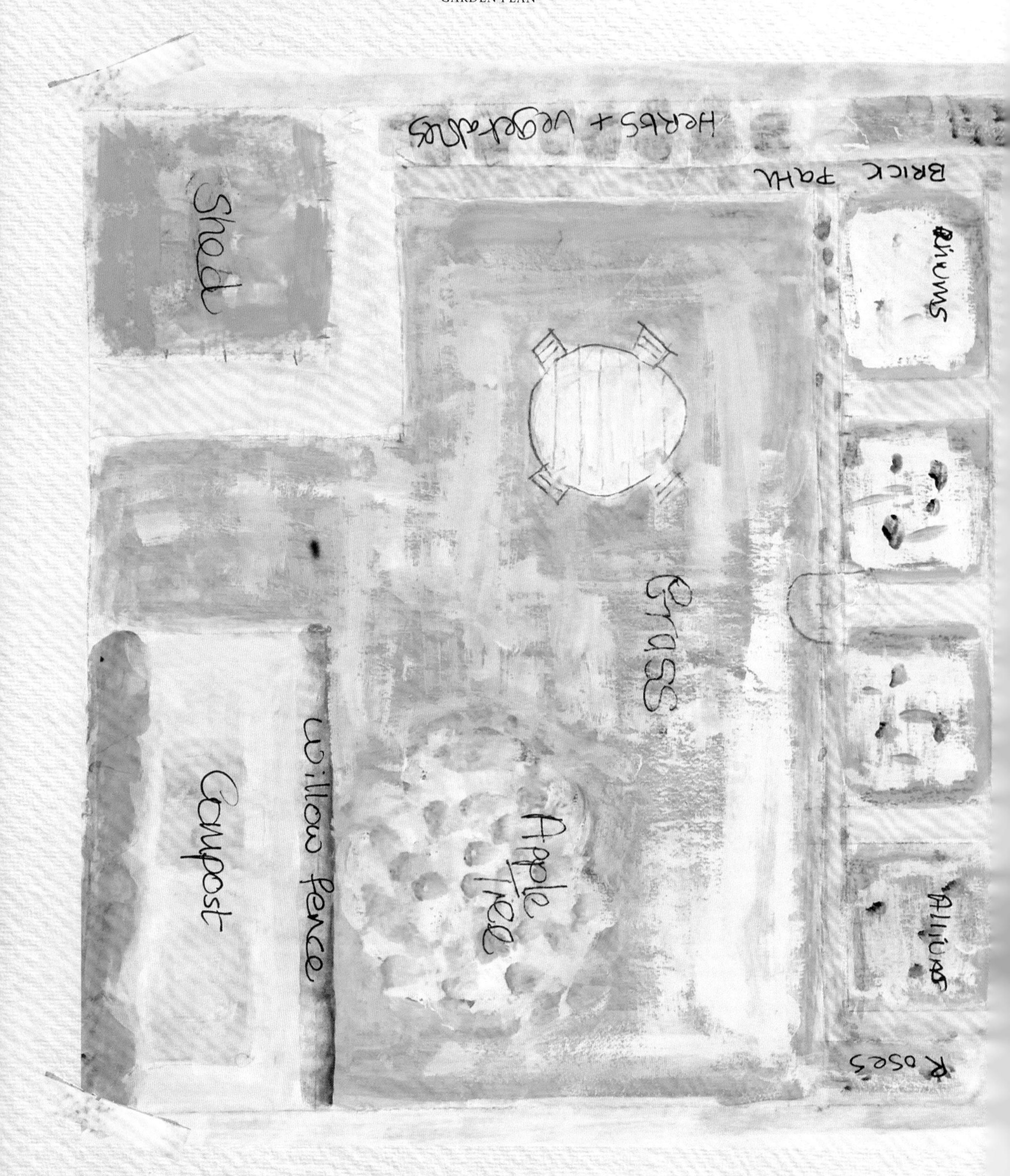
Herbs + vegetables
BRICK PATH
Shed
Alliums
Grass
Willow fence
Compost
Apple Tree
Alliums
Roses

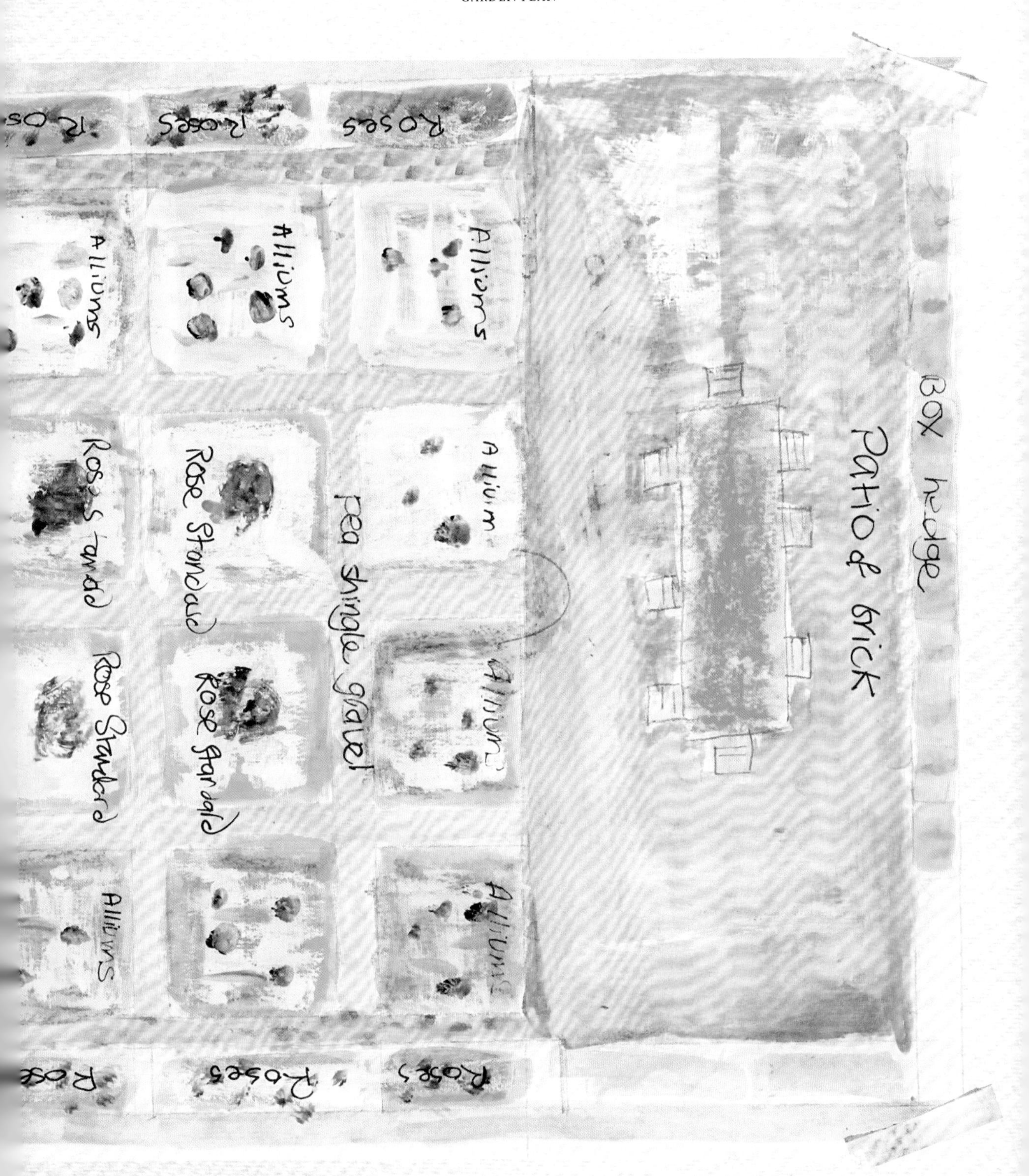
Box hedge
Patio of brick
Pea shingle gravel
Roses
Roses
Alliums
Alliums
Alliums
Allium
Rose Standard
Rose Standard
Rose Standard
Rose Standard
Alliums
Alliums
Alliums
Roses
Roses